Havoc

Edge Of Oblivion

By

S.A. Coons

Special Thanks To

Special thanks to everyone that helped make this book what it is.

Bryce Biskup

Joshua Dennis

Shannon Kunz

Frederick "Fred Nasty" Nash

Alyssa San Pedro

Seth Simpson

Last but certainly not least, my family who supports me no matter what.

Kasey Coons

Charley Coons

Cindy Wood

Contents

Chapter 1

James Minder sat in his small ship, staring into the vast emptiness of interstellar space. An unremarkable looking man by all measures, average height and build, brown eyes, and light grey brown hair. At 72 years old, he was considered middle aged in the 25th century there too. He was just average. He could not be picked out of a crowd except by those who knew him, and those were few and far between. But none of that mattered here, he was lightyears from the next living being, searching for the key component to make his outlandish plan come to fruition.

It had been three weeks since he finalized the deal and transferred all of his funds, both borrowed and owned, into an anonymous account in exchange for the six *highly illegal* XS-12 Core Busters. The XS-12 was developed during some long-forgotten war and was used as the ultimate weapon. With a single XS-12, an entire planet could be destroyed from the inside by creating a short-lived singularity injected into the target planets core. After it was used with great effect by the winning side, it was outlawed immediately after the war concluded. All remaining devices and data were supposedly destroyed due to the public outcry. James did not care about the public and did not even know who won that war or what the war was about. What he did know was that even having working knowledge of the Core Busters could result in a summary execution by any law enforcement or military in all of the controlled space.

Suddenly, there was the flash he had been waiting for. An Alcubierre drive (Que-Drive) equipped ship popped into existence. A second flash less than a minute later could only mean two things. The first was that a second ship jumped in, which would signify that James had been double crossed and his life would soon come to an abrupt halt. The second and more preferable explanation was the ship that jumped in, dropped off his cargo, and jumped out again.

The flash was a light minute away, and if there were two ships, he would pick them up on his passive sensor scans within the next couple of seconds, especially if they were running active scans. James had to remind himself to blink as he sat in the pilot seat, his eyes affixed to the sensor monitor. After what seemed like an eternity, a pixilated form of a single ship appeared on the screen and then disappeared again, leaving just four pixels in the deep black.

James was mesmerized by the enormity of what those four tiny pixels represented, his grand scheme, and they were so close. But, he reminded himself that there was still so much room for error and that he must tread lightly. He waited another two days in his ship before he planned on moving toward his prize. He passed the time by checking on his old recon ship systems. The ships design was still in use by the outer fringes as well as private couriers who operated in the core worlds, mostly because they were so cheap, robust, plentiful, and could be run by a crew of one or two with minimal training.

The system checks turned up with nothing out of the ordinary, so James turned up the artificial gravity and started working out in the cargo area. When he reached his arms out to his sides, he almost touched both sides of the cargo bay. So, he was limited in the exercises he could perform, but he managed to work himself up to the point of exhaustion anyway. Then, he laid sleeplessly on his bunk, planning his next course of action.

After the self-imposed precaution period was over James throttled up his ships mass reaction engines and burned toward his prize. He knew at any second the flash of a que-drive could mean the end of all his plans, as well as his life. The little over eleven-million-mile trip took the better part of the day with the last half of the trip spent in the counter burn with the inertia dampers maxed out so that he could match the Core Buster's delta-V. Once he was within the magnetic grappling range, James launched the grapple line on the lowest speed setting to ensure that his volatile cargo was not subject to any unnecessary shocks.

Once the Core Buster's reached the hull of his ship, James donned his hardened black and orange EVA suit and began to depressurize the cargo area. Once the vacuum was achieved, James opened the outer hatch, affixed his safety cable, and stepped into the black. The nausea was almost immediate as he went from a standard one earth gravity (g) to the sensation of free fall when he passed the threshold of the outer hatch. Once he regained control over his stomach, he mentally chastised himself for not turning off the gravity before exiting his shuttle. Now, he would have to go back in and turn it off. He vowed no more rookie mistakes.

James pulled himself back into the shuttle with some force to come into the artificial gravity all at once. However, he did not orientate himself correctly. In his moment of vertigo, he lost track of which surface was the ships floor. Instead of coming in with his feet facing the ships deck, he came in with boots facing the back wall and promptly slammed on to his left side when he entered the ship, only avoiding injury because of his hardened EVA suit. This time, the tirade about his own mental competence was on full display for no one to see. Curses about all of the old gods, the new gods, and some deities he made up on the spot flowed from him like the rivers of old earth for several minutes, that is, after his diaphragm stopped spasming.

Once James pulled himself together, he promptly looked at the ships deck and finally understood why someone had painted yellow footprints on the deck. Once upon a time, they had made the same mistake he had, and they made sure that they would not repeat that mistake. Stomping over to the gravity generator control panel, he slowly lowered setting to .1g's. James now weighed 17 pounds, but had to be careful not to move too fast or push off the floor too hard, or he would end up with his third rookie mistake in less than an hour.

After the recent mistakes, James double checked his safety cable and made sure to find the footprints before stepping outside again. This time, the transition from gravity to no gravity was expected and much less abrupt. He spotted his prize connected to the end of the grappling cable and begin to slowly wench them in. As he readied to cross the threshold back into the ship, he ensured that his cargo and himself were oriented with the inconspicuous footprints.

James secured the cargo and turned up the gravity to a full one g. He stripped off and stowed the EVA suit, handling it with a bit more care than he ever had before, thankful that it prevented him from hurting himself. He turned back to the cargo area and was stopped in his tracks. He had done it. He had obtained the last six XS-12's known to man. Sitting there, *in his ship*, was a neat three by two stack of metal crates with enough raw destructive power to bring all of the known space to its knees, and he was determined to do just that.

Chapter 2

After several minutes of pure euphoria, reality came crashing in. The enormity of what he was trying to do began to weigh on him with a force he had never before experienced. Sitting before him was the power to reshape all of human space, but this type of power could only be used for evil. Nothing with such indiscriminate destructive power could have any noble purpose.

James thought about all the events that led up to this point; the abject loneliness of his life, the one in a billion chance of winning his system's largest lottery, and the isolation he was forced into while avoiding the people who were trying to get at his money. All of the research, backdoor deals, all of the broken lives he left in his wake to get to this point. The only reason he was able to get the Core Busters was because, after centuries of sitting in their cases, the arms dealer that previously owned them did not think they were authentic, and if they were, they most likely no longer functioned. The price for the six XS-12's was still in the billions of credits, James had to go into debt to procure them and the other components to carry out his plan.

But now that he had everything, could he really do it? Could he really kill that many people for no other reason than his vanity? The doubt began to creep in, and a cold chill ran through his entire body, but he shook himself out of his sorry state. He had already sacrificed so much to get here.

He could always change is mind later, he just needed to see if they were the real thing.

James had long ago selected his home planet, Jin, for the test. It was the place that made him into the thing that he was. He made his way into the cockpit, and during the short walk, he decided that the XS-12's were probably fake, so what was the harm in trying? Once seated in the pilot seat, he selected the planet Jin as his destination and engaged his Que-drive.

The trip in null space would take a full week and James passed the time by studying the hardened operations tablet that came with each Core Buster. The more James read, the more he started to believe these might be the real thing. The operation of the XS-12 was fairly straight forward: Place the whole container within five miles of a fault line. Enter the 24-digit code that he had. Pass the biometric scans, which were disabled. Then, select the mode of deployment. One of the choices for the weapon deployment was a bit odd to James, but maybe it could be useful.

The first option was the immediate deployment of the weapon. This started the laser bore as soon as the selection was made. Depending on the mass and make up of the planet, the XS-12 would light its plasma torch and drill to the core in approximately 90 minutes plus or minus 10 minutes and start its detonation sequence. Thirty seconds after that, the planet would begin to implode to the point where the matter could no longer be compressed. Then, a miniature Big Bang would manifest where the planets core once was. No planet in the forgotten war had survived this attack.

The second selection was almost identical to the first, except for a time delay function with the option to delay up to 100 years. The third option was a dead-man's switch. Once the 24-digit code was entered and the icon was pressed, it could not be released without entering a 52-digit abort code, which he did *not* have. If released without entering the code, the singularity contained within the munition would break containment and explode in thirty seconds. A warning in the manual stated: "Without the sufficient mass of a planet surrounding the XS-12, the XS-12 may not produce an explosion large enough to completely destroy a planet". James thought out loud, "So, the less mass the singularity has surrounding it, the smaller the explosion." He then asked the tablet, "what would happen if the XS-12 exploded in space?" The tablet did not reply, nor did it contain the answer in its data.

The forth option was a remote detonate function that allowed the user to choose a function similar to option one. Once the 24-digit code entered, the laser drill would activate, and the XS-12 would complete its cycle. A sub-option under this section was the XS-12 detonating 30 seconds after the remote was triggered. The distance the XS-12 could be remotely triggered was unlimited due to its quantum entanglement with the hardened tablet. This option also came with the same warning: "Without the sufficient mass of a planet surrounding the XS-12, the XS-12 may not produce an explosion large enough to completely destroy a planet".

The fifth and most interesting of the functions was a variant of option two. However, the big difference was a randomizer. The user could choose a no earlier and no later time up to 100 years into the future. Then, at random between the two times, the XS-12 would arm and complete its drill and detonation cycle. James thought about this function for quite some time but could not come up with a logical reason why this was an option.

Each munition had its own unique 24-digit code that James tried to memorize. However, by the time he memorized the second code, he forgot the first. So, he went with a different approach. He decided to label the mat green XS-12's and their quantum entangled tablets in sequential order: 8, 9, 10, 11, 12, and 13. After all, the designers of the Core Buster were not the only ones that could be whimsical. Then, he memorized the code to Lucky Number 11.

Once James had the code memorized, he began to recite it while he was carrying about his normal daily routines. The biggest change to his routine was the fact he now had to share his workout area with the most powerful weapons ever devised. He did feel a little awkward using his newly purchased multi-billion credit gym apparatuses. However, as he worked out, he continued to recite the code as if it were his new mantra. He did not know if he would need to do that, but he did not intend on taking the codes with him when he set out with his plan. By the end of the week, James knew the code and could recite it from a dead sleep if required.

While traveling, the doubts began to chip away at his resolve. Was this what he wanted to be remembered for? Could he really murder over a billion people that he did not even know? When those thoughts occurred, he forced himself to remember all of the neglect he endured at the hands of his parents and his class mates and how all of those people came rushing back into his life when he won. "That had to be a large enough sample size to condemn the planet and the rest of humanity, right?" he asked, but no one answered.

As he dropped out of null space, his home world came into view. He was both awed and disgusted by it. Jin: the place of so many terrible memories. James felt his resolve stiffen. He could do this, he would make the universe recognize him. While he continued on his approach, the messages that had been waiting on him to get within range started pouring in. They came from all walks of life, but they all wanted the same thing: his money. There was no message from a loved one because he had none in his life. James smiled to himself because they did not know he was almost as penniless as they were. He had just enough credits to cover the refuel and refit of his ship. Then he would be just like them, broke. He did not answer a single message, and after listening to the first few, he just deleted all of them.

Chapter 3

James was by no means a great pilot, he relied heavily on his ships internal systems to do most of the work. The only thing he really had to do was tell the Jin spaceport controller what his intentions where. That way, the automated system could route him to the correct landing spot. Once he told the controller that he was looking to refuel and replenish his food stocks, he was given an option of landing zones based on the cost. James had just enough to splurge and go for the high-end spot with all the amenities that he could want. However, he did not want to attract too much attention and definitely did not want a cleaning crew in or around his ship. He would take care of that himself. James landed with out any issues and was far enough away from everyone else to put his mind at ease. He disembarked his ship and entered the lock down codes.

James had a full 27-hour day to kill before his ship would be ready. Since it was just after sunrise and he saved quite a bit of money on his landing fee, he was going to walk around and try to convince himself one way or the other what the next day would bring. It was an odd feeling; looking at all the people and places, knowing that whenever he chose, they would all be gone. As he passed some of the familiar streets of his home, it jogged memories, and they continued to flash before his mind's eye, almost all of them bad.

James knew he did not have a horrible childhood. His two older brothers and younger sister all lived normal lives.

They did not reach out to contact him for years, it was as if he was forgotten. That is until he won the lottery. The day his name was announced as the winner, his entire family decided then would be a good time to try to reconnect. At first, James basked in the attention that he craved. But no matter who he was talking to, the conversation always turned on how James should "loan" them money. After a couple of months, everyone demanded that he give them the money *because he owed it to them.* Shortly after that change in conversation, James cut off all contact with everyone and became a recluse. James wanted attention and to be remembered, but not for being a sucker that gave all his credits away.

He wondered aimlessly around the busy city, trying to decide one way or another when he heard what sounded like an animal crying along with men laughing and cheering. James quickly turned into the narrow alley to find four men in what looked like their early twenties surrounding a small brown puppy. The young men laughed as the little dog limped around and fell over because one of its back legs was twisted at an unnatural angle. "Kick it again!" the largest of the group yelled. He was the only person whose face James could see. The expression of the fully-grown man was something James would never forget, pure joy mixed with excitement. *Evil.* James could almost feel the blow as the guy kicked the puppy into the side of a couple of trash cans. The puppy's cries of pain and anguish broke something in James' soul. When the trash can fell on top of the little brown dog, it cried no more.

James charged the thirty feet separating him and the largest of his new-found enemies, he then launched himself feet first at the small of the largest man's back. The only person that saw James charging them shouted at the group, "LOOK OUT!" The big man turned with surprising quickness, leaving James sailing in the air past him, barely making contact. James landed hard on his side, but managed not to lose his breath or bounce his head off the concrete. As James rolled to his back, he looked up and realized he made a tactical error. Staring back at him were four sets of eyes that projected the violence they were about to rain down on him.

Before James even moved, the first blow landed. Coming hard from his right side, he felt the toe of a boot crack one of his ribs. The pain rippled through him, causing his vision to dim briefly. The next kick landed on his stomach and brought reality back into a sharp focus. Out of pure desperation, James lashed out with all his might; his leg could manage and connected a solid blow with the man's knee who kicked the puppy. James felt a sickening wet crunch. The man crumpled to the ground, screaming and holding the afflicted leg. With the man at his feet, disabled, James hoped that the others had their fill and would take their comrade and leave. He was wrong. The largest man grabbed James by his hair and pulled him to his feet. "You're going to bleed for what you did to Hunter.", he spat as he drew a small pocket knife and pressed it into James' left cheek. James was in shock as the knife sank into his flesh and made a shallow incision from his cheek bone to his chin. Blood poured from his face as if a faucet had been turned on. It happened so fast he did not have time to scream or fight back. After the knife was removed from his face, the largest man stepped aside just enough for a fist to connect with his already battered torso, knocking the wind from his lungs and his breakfast from his stomach. “Maybe next time you’ll mind your own business, *hero.*” The large man said as he spat and kicked James one more time for good measure. Two of the men grabbed Hunter by his arms and helped him up, but James barely noticed, his vision started to tunnel and then go black.

The blackness was peaceful. The blackness was painless. Then, the blackness was interrupted by something touching his face. Something soft... something wet. All of a sudden, reality snapped back into place and the pain made him bolt upright. He understood that had been a mistake as added pain caused spasms, which only caused more pain. James cried out and almost blacked out again. The only thing tethering him to reality was a small brown puppy whimpering between a trash can and the building he was formally pinned against. The small dog must have been licking his face to wake him up. Two golden eyes seemed to gaze into his very soul, and the bright clear eyes were in such contrast to the rest of the broken animal that James temporarily forgot that he was also hurt.

After what seemed like an eternity, James managed to get close enough to the puppy to gently stroke his head and try to gain his trust. James knew that the whole “fight” only lasted a minute at most, but it was almost noon before he was able to pick up the chocolate colored puppy with the piercing golden eyes. The pup let out a quick yelp once James picked him up and had recovered enough to try to walk. As he held the puppy in his arms, James could feel him shaking and hear the wheeze with every breath the pup took.

James pulled out his comm and looked up the nearest vet. Luckily, it was only a few blocks away. James stepped out of the alley, went back onto the main street, and was shocked. As soon as he was on the sidewalk, he saw that he was surrounded by people. How many people had walked by him and saw what happened? They all looked the other way. They walked past him, seeing that he was covered in blood, some went as far as to step onto the street to avoid him. James' heart hardened.

As quickly as his broken and battered form would allow, he made his way to the vet's office. After the short trip, he stepped into the foyer of the office and felt the room go silent. One of the receptionists grabbed her comm and started to dial what James could only assume was the police. He slowly approached the counter and cleared his throat, "Excuse me, but I have a puppy here who needs medical attention." Only then did the receptionists notice a small dog curled up in his arms. "I'm sorry, Sir. But we are not a charity clinic. You will have to take that dog somewhere else." she stated with her hands on her hips. James realized that she thought he was homeless. "I assure you... Jenny," he said as he read her name tag, "I have the funds to pay. Check my chip." He held out his hand. Jenny ran the scanner over his hand as he awkwardly held the puppy. After reading the screen on the scanner, she looked up at him, "We'll be able to help you, Sir." James looked at the puppy, "What's next?" The woman gave him a withering look, "I will have to try fitting you into the schedule somewhere. How does next week sound?"

"Next week?" James asked through gritted teeth. "This little guy has a broken leg, and I am sure, some type of internal bleeding. I don't think next week is an option."

"We can see him today," the receptionist said, looking slightly more interested. "But you will have to pay the emergency fee."

"That's fine, I'll pay." James nearly spat the words at her.

"Okay. Well, what's the little guy's name?" The receptionist asked in a sickeningly sweet voice.

"I found him in an alley after some guys used him as a kick ball," James noticed how once she knew he had money and was willing to spend it, her attitude changed.

"So, this isn't your dog?" she asked. "No, I think he's a stray."

"Well Sir, it would be cheaper if we just put him down if he is in as bad of shape as you think he is in. But that will cost you 40 credits." The sweetness disappeared out of her voice.

"Why would I have brought him to your office just to pay money to have him put down?" James growled. "Look, I'm just trying to save you money, Sir." Jenny snapped.

"Look, he's my dog now, and I want him treated... please."

"Okay then, what is his name?" Jenny asked the sweetness back in her voice.

"...... Loki. His name is Loki."

"Okay then, please give Loki to me. He should be as good as new by this afternoon. Now, then, go to the restroom to clean yourself up. You are frightening the other pet parents.

Chapter 4

James left the veterinarian's and thought perhaps it was time for him to take care of himself. He knew where the nearest hospital was. He had been there on several occasions while growing up. New George hospital was across town, and James had to hail a cab to get there. As James comm'ed the cab service, he was notified that the next cab would be there in 2 minutes. With just a few moments to kill and his comm already in hand, James began to scroll through his contacts, looking to see if there was anyone worth talking to. He found no one.

When the cab pulled up, James gingerly lowered himself into the back seat. The cab already knew his destination and slowly rose to its cruising height of 10 feet and accelerated down the street. James turned off the chat function in the automated cab and began to think about what just happened. Somehow, through the course of the morning, he tried to be a *hero*, got beaten pretty badly for that act, became a pet owner, and was now headed to a hospital. James was feeling better about his decision to wipe this planet from existence with every pained breath he took.

The hospital was no adventure, and he was back on the street in just over an hour with a skin wielded cheek, nanite and calcium injection for his ribs, and a stern warning not to do anything stupid for the rest of the afternoon. James hated nanite injections they made the affected body part itch like crazy, but it was better than being scared to take a breath, he supposed. Now, he needed to find some clothes that were not covered in blood and torn to shreds.

The auto-fab shops were just a couple of blocks away, so James decided to walk. All though, he promised himself no more heroics after. After all, wasn't he about to become the most feared villain in the known galaxy? Due to it being later in the afternoon, most people were starting to get out of work, and those that had kids were going to the local school to pick them up. As James passed by the school, he noticed a large range of the human condition; children happy to see their parents, while others were being scolded. He thought about how, in less than 14 hours, all of this could be a floating cloud of detritus, a marker for the rest of the galaxy to remember him by. His iron resolve began to soften when he saw the joy on some of the families faces. After being separated for only a couple of hours and for something that happened seven days out the ten-day week, how could you miss each other that much? His family never acted like that toward each other. Why should these people enjoy their lives when he never got to experience that? His resolve hardened.

James continued his journey to the section of town that was the hub for the clothing merchants. He entered one of the more modestly priced retailers and was body scanned as he walked in. All of the displays turned into pictures of him wearing the clothes that that particular shop was selling. James scanned through the displays until he found a fair priced and practical two-piece jump suit that was common for spacers, which he now considered himself to be. All he had to do was select the colors and patterns. He chose black with orange accents while he approached the bored looking attendant. He had to get her attention by clearing his throat, just so she would acknowledge him.

She stared at him for a few moments before she spoke, "We are not a charity. I am going to have to ask you to leave." "Why does everyone keep thinking I'm homeless?" James inquired. "Maybe because you look and smell like you live in garbage," the bored attendant sniped. "Well, I am here to rectify that particular situation," he said, keeping his tone neutral.

"Fine, let me scan your hand," she said while waving her scanner at him. James held out his hand once again, as she looked over his account. She simply asked, "How many sets do you want?" "I'll take three, if it's not too much trouble," he replied. "That will be 50 credits, and you can pick them up in two hours. Or you can pay an expedited fee, and I'll move you to the front of the queue for our fabricators."

James looked over her shoulder and noticed that none of the fabricators were printing any clothes. Through gritted teeth, James simply stated, "I'll take the expedited option."

"Okay, that will be 60 credits and will take ten minutes." She held out the scanner for payment. James placed his hand under the scanner, and the credits where deducted from his rapidly dwindling account.

"I suggest you use the restroom in the back to wash some of the grime off. You don't want to ruin your brand-new clothes. As a matter of fact, if you toss your old clothes in the trash back there, I will throw your first set of clothes off the printer back there for you."

"I would like that. Thanks," James said as he moved toward the restroom in the back. He locked the door to the restroom, began to undress, and use the sink to try scrubbing off some of the grime and blood that the hospital missed, which was a lot. He ended up removing all his tattered clothing, even his under garments were ruined. He then tore up his old clothes to make rags and gave himself a sink shower to the best of his ability. About half way through, the attendant knocked on the door, waited a split second, then unlocked it from the outside. James stood there as the attendant walked in and saw James covering himself with a torn piece of his old jump suit. "I waved the expedited charge and used the extra credits to print some underwear for you. Looks like I made the right call," She smirked and threw his new clothes on the floor next to him.

"Thanks," He said while waiting for her to leave, but she did not. Having enough, James dropped the piece of the jumpsuit and stood there, hands on his hips, until the attendant smirked and left. "What the hell was that about?" James asked the universe while getting dressed quickly with the door wide open, because what was the point. When James exited the restroom, the attendant had resumed her position at the counter, not even bothering to look at him. She just pointed to the two sets of clothes stacked neatly in a bag. He walked out without saying another word.

James checked the time and noticed a missed message. Loki was ready to be picked up and the vet had some questions for him. "Great, today just gets better and better." James hailed a cab while still thinking about the strange encounter with the clothes shop attendant. "She just saw me completely naked, and I don't even get her name." James climbed into the cab and headed back to the veterinarian's clinic. Still very confused about what just happened, he had completely forgotten about what he was planning for tomorrow.

Chapter 5

James exited the cab with ten credits deducted from his account. While he was not completely broke yet, he was reaching the point where money was tight, and he still had to figure out what will happen after tomorrow. He walked through the foyer for the second time that day. This time, without the stares and hush spoken tones about him. As he approached the counter, he was glad to see that a different receptionist greeted him, but her name was also Jenny. This was either an odd coincidence or something else going on. "Good evening, how can I help you?" *So-called Jenny* asked with the same fake sweetness as the other. James wondered if it was a training requirement that gave the two receptionists the exact tone and name for that matter. Jenny loudly cleared her throat and snapped James out of his impromptu ponderings. "Sorry... *Jenny*, I am here to get Loki," James stammered.

"You must be Mr. Minder. Loki is ready to be picked up, but our veterinarian wants to talk to you. If that's okay?"

"If it's all the same to you, I would prefer to collect my dog and be on my way."

"I am afraid that Dr. Robinson was very insistent on meeting with you before you were able to leave with Loki," Jenny said sternly.

"Well then, it seems I have little choice in the matter," James huffed.

"Okay, follow me, please." Jenny turned and held open a door that led further into the clinic. As James followed, he was quite surprised at how large the clinic was. There was no way he would find his way out without some help. Jenny finally stopped at a door that looked like all the rest, in a hall that looked like every other hallway they had been through. James could not be entirely sure that they had not walked in circles. Jenny knocked. "Dr. Robinson, are you in?"

"Come in," a deep, gravelly male voice answered. Jenny pushed open the door, "Mr. Minder is here to see you."

"Great, send him in. Could you grab Loki and bring him here as well?" Dr. Robinson asked.

"Of course, Doctor." Jenny let James into the small office. Dr. Robinson stood and shook James' hand, then pointed him to an uncomfortable looking chair. "Please have a seat," he said. As James sat, he noticed that the doctor was shorter than him, but was almost twice as wide and had solid muscle. The doctor looked younger and had flaming red, close-cut hair. James absently wondered if he worked for his physique or paid a body shop to keep him in that condition. "Mr. Minder, I will cut straight to the heart of the matter. I need to know where you got Loki from." He said with an expression that left no room for doubt that the doctor was not in the mood to beat around the bush. Unperturbed, James answered, "I already told *Jenny,*" holding finger quotes up as he said the receptionist name, "Ask her."

"I did ask her, and now I am asking you." The doctor said while leaning over his desk.

"Then I will tell you the same thing I told her, which, by the way, happens to be the truth. I saw a couple of men kicking him around and intervened. For that act of heroism, I received a broken rib and had to get my face lasered back together. Are we done? I'd like to be on my way." James nearly spat his words out. Jenny chose that moment to produce Loki and promptly sat the puppy in James' lap.

The little dog had an instant calming effect on James, and he appeared to have the same effect on the doctor. James studied his new companion and hardly recognized the little dog. He was clean and happy, lazily wagging just the tip of his tail as he yawned. Loki turned in tight circles on James' lap, promptly flopped down to his belly, closing his big golden eyes, and going to sleep.

"Mr. Minder, let me try this another way. Do you know how dogs got to this planet?" The doctor asked.

James thought for a second. "I have never thought about it. This," he said as he nodded toward Loki, "is my first interaction with a dog outside a petting zoo."

"Well, that is because dogs are not naturally occurring on Jin. They have to be specially ordered, and then they are grown in an artificial womb. The whole process takes about a week," Dr. Robinson stated, "And all dogs are born sterile and are registered to the person that placed the order."

"Okay, what does that mean for Loki?" James questioned, worry starting to creep its way into his voice.

"Nothing bad, I assure you." Dr. Robinson noticed the change in James' tone. "Let me explain. Loki has a highly specialized DNA sequence inserted into his base DNA code, while that is not unheard of, it is very expensive."

"That means you know who ordered Loki, right?" James asked, not really wanting the answer. After all, he had taken a beating for the pup.

"That's one of the reasons I wanted to talk to you. Loki's owner died before she was able to get him. Normally, when an owner dies before they can pick up their pet, the dog is terminated," The doctor stated, "I do not know how Loki got away from the lab, but I deleted his old registration and registered him to you."

James let out a breath he just noticed he was holding and simply said, "Thank you."

"The registration was the easy part. I just wanted to make sure that you did nothing nefarious to obtain Loki. After that, I wanted to talk about the nature of his DNA modifications. Let me ask you a question, Mr. Minder. Are you a spacer?"

"Yes. How did you know? And what does that have to do with the current situation?" James asked.

"Your suit. Your suit is the answer to the first part of your question, and the second part is that Loki was designed to thrive in space. He will do well on a planet. But in space, he will thrive."

"Could you be a little more specific on thrive in space?" James queried.

"Loki's original owner was a long-haul spacer that ordered Loki to be her traveling companion. That is why

Loki has certain adaptions that make him comfortable in zero g environments; such as tiny weak suction cups on his paws that don't do anything in gravity but in zero g, he will be able to walk on the floor, walls, and even the ceiling. He is able to withstand a hard vacuum for two to three minutes before he suffers any ill effects. There other small adaptions as well, but the most interesting are not related to space travel."

"They must be pretty amazing adaptions if they beat surviving hard vacuum," James stated in disbelief.

"Oh, they are. The first being that he will have the intelligence of a five-year-old child, a smart five-year-old" Dr. Robinson stated, "He will also be able to solve simple problems and understand most speech. That is, of course, if you train him correctly."

James was shocked. "Is...is there anything else I should know?"

"One other interesting adaption. Loki has the ability to influence his owner's mood, as well as others around him."

"I'm sorry, are you telling me not only do I have a super dog that can survive vacuum, is extremely smart, and he can also control my mood?" James asked, feeling like he should be more upset than he was.

"No, Loki cannot control your mood. He can smell when your body produces a spike in cortisol. You know, the stress hormone. Then, he can release his own pheromones to counteract them. With the right training, you can teach him not to do that or to produce more on command. However, ask yourself how your mood is now compared to when you walked in."

"I feel better than when I walked in, but that could be because you have not hit me with the bill yet," James said.

"Look, I am an animal lover. That is why I chose this line of work. I saw the footage of you when you walked in. You did indeed take a beating for that little dog, so I am going to cover the cost of his bill today. I am also going to provide you with the puppy starter pack that all my clients get when they order their dogs through me. It has the food synthesizer recipe for all of his nutritional needs, as well as several other items you should find useful. Now to the last order of business, what color collar and leash do you want? "The doctor asked with a smile on his face.

Chapter 6

James exited the vet's clinic with Loki in lock step. The puppy was wearing his brand-new black collar attached to an orange leash. The puppy pack that the clinic supplied was more of a crate, and James had to have it delivered to his ship. He would have to go straight back to his ship to beat it. He definitely did not want anyone in or around his ship without him there. James comm'ed the cab, and less than a minute later, he was on his way to the space port with Loki refusing to sit on the seat next to him. Instead, once again, Loki plopped down in his lap and fell asleep.

James had to carry Loki the quarter mile from the terminal to his ship, not because he was worried about the pup running around an active spaceport tarmac, but because when James got out of the cab and set Loki on the ground, the puppy promptly laid down and refused to move. As James switched the pup from arm to arm, he thought about all the events that had happened that led him to carry the little dog around. First, he was beaten and had his face sliced up, then almost got into a shouting match with a receptionist at the vet clinic, was gawked at while he stood completely naked, and, to top it all, he was now the owner of a super dog and received over 1000 credits in free dog care from the vet's office.

The weight of the dog pulled on James' shoulders and back as he approached his ship. James looked down into the golden eyes of the pup and said, "How are you so heavy? This the last time you get a free ride." As if the little dog completely understood, Loki started wiggling in James arms until James had to set him down. Loki then started pulling on the leash, heading straight toward his new home. "HEY! Loki, stop pulling." Loki looked back, lowered his head, tucked his tail, and then slowed his pace. "Sorry buddy, I need time to adjust to having a dog. I'm not sure how all of this works." James said apologetically. Loki picked up his head and once again got into lock step with James, seeming to accept it. "I wonder how you know which ship is mine?" James wondered aloud. Loki turned his head up to look at James, drew in a huge breath, and blew it out as loudly as possible. "Smart dog," James said while noticing Loki begin to have just a little more bounce in his step as he happily wagged his tail.

When the pair reached their ship, James opened the side hatch, and Loki ran in as if he already had lived there for years. As James stood at the hatch of his ship watching Loki run the main corridor from the bow of the ship to the stern of the ship, he wondered where all that energy came from. A few minutes ago, Loki was so tired he couldn't walk.

"Excuse me, sir." A delivery woman said, startling James. "Are you James Minder?"

"Um....yes, I am." James stammered. "Are you sure?" she asked.

"Yes, you just caught me by surprise," James replied.

Having heard another voice, Loki came up to the hatch and barked once at the intruder. He then proceeded to run up to her and jump at her legs, wanting to be picked up. The woman set the hover dolly carrying the puppy crate to land and bent down to pick up Loki. "Oh, my goodness! You have a puppy!" She exclaimed while snuggling the chocolate colored pup. "Yeah, and he is going to be a fine guard dog from the looks of it."

"Well, he sure stopped me," She giggled while Loki wiggled up to her face to get better licks. James chuckled, "Well, I guess he did."

After a few more minutes of pure bliss, Loki was calm enough to be set down. She gently set Loki on the ground. As soon as his paws hit the tarmac, he promptly sprinted full speed back into the small space craft and disappeared inside. "I apologize about that. I got him today, and apparently, we have to work on manners," James said sheepishly. "Don't you dare," the delivery woman said with an overexaggerated huff, "If I am not greeted like that every time I see that little dog, I will be very disappointed? By the way, what's his name? It is a he, right?"

"His name is Loki, and I guess I will have to make an exception for you when it comes to his behavior." James smirked, "Do you have a name?" "My name is Alyssa," she said offhandedly, she turned her attention back to Loki. "I have never seen a puppy in real life. That is going to make my kids so jealous. Can I run in and get a vid with him?" James almost said yes, and then a cold shiver went down his spine as he remembered what was occupying his cargo bay. "I'm sorry, but I can't let you in there... but I can bring him out again. And while you are getting your vid, I'll load the cargo into my ship." James said quickly. "Boy, you really don't want me to go in there, do you?" The delivery woman asked with a little bit of curiosity. "It's not that. It's that Loki is a puppy and is not potty-trained. Plus, to tell the truth, I was not planning on having a dog this morning when I left. Sooo, I have some work to do in there." He said while pointing to the crate on the hover dolly. "Well, that makes my job easier," She said while grabbing her come out of her pocket. "Now go get Loki, I have some kids to make jealous." She said in a playful tone.

Alyssa, left when she got a comm asking why she was not on her way to her next delivery. She quickly made up a story about a problem with the current delivery and said she had just fixed it. She quietly handed Loki over to James and mouthed the words, *thank you*. An hour later, James had set up all of the items that were included in the puppy box. James noted that the doctor had not skimped out on the puppy starter pack at all. Everything from waste disposal, food recipes, zero g water bowl, and toys were there. Loki would want for nothing, it seemed.

The day was coming to an end, and James laid in his small bunk, looking at the ceiling. "Today was pretty wild. Huh, bud?" James asked Loki. The only response he got was two little wags of just the tip of Loki's tail. "You're no help." James was not expecting an answer but was surprised at how easily he adapted to speaking to the new addition in his life. The day had left him torn and now he was less sure about how he felt regarding his grand plans. The whole purpose of the day was to solidify his plans one way or the other. Now, laying in his bunk, he could not stop from changing his mind over and over again. His thoughts were interrupted when he heard a soft whine next to his bed, "Loki, the vet said you should sleep in your own bed. After all, it does take up a good chunk of the usable floor space in my ship." The whining stopped for a few seconds before beginning again with a little more insistents. "FINE, you can sleep up here with me but have to figure-" Loki was now laying on his chest, "-out how to get up here." James mumbled, finishing his sentence.

Loki's calming presence did help James finally go to sleep after a while, but only after he decided that he would fly out to the fault the next day. After the decision was made and he adjusted Loki to a more comfortable position, he was finally able to drift to sleep listening to the little puppy snore softly.

Chapter 7

James' sleep was troubled, and more than once, when he woke up, he was wearing a puppy hat. Each time he moved Loki to the foot of the bunk, the little dog would find his way back up toward James' head. That did not bother James nearly as much as the hellish nightmares. Fire, intense pain, the sensation of being pulled apart, and then just black with the sensation of falling. Each time, he jerked awake in a cold sweat. When he would fall back asleep, the cycle started again. The only reason he was able to go back to sleep each time was the rhythmic snore of Loki.

As the sun rose over the horizon, it began to light up the spacecraft, and James had had enough of pretending to sleep. He slipped out from under his puppy hat, trying not to wake the little dog. His cabin was just big enough to hold his bunk, a small work station, a waste recycler (now a dog waste recycler, as well), and sonic cleanser just large enough for him to stand in and clean himself. He left his cabin and stood in the main corridor. If he went to his left, he would enter the cockpit. If he turned to his right, he would end up in the small cargo area. He turned right, walked past the common area/kitchen, and stood just outside of the cargo area. Normally, this door was closed when the craft was in operation. However, because the ship was just parked on the tarmac, all the interior hatches were open.

There before him sat the most powerful and destructive weapons man had ever devised. He sat on the deck and stared at the XS-12's as though he was seeing them for the first time; such a simple looking green case to be able to completely destroy a planet seemed at odds with reality. In true military fashion, the only marking on the case was an embossed arrow with stenciled type spelling out *THIS END UP*. James sat in the entrance way to the cargo area and thought about what he was about to do. The 24-digit code flashed through his mind as his eyes came to rest on Lucky Number 11. James knew at that moment that he had to execute his plan, too much time and money was spent on getting the XS-12's not to at least try them out.

James heard the waste recycler whirl to life and was startled, thinking that someone else was in his ship with him. After a few seconds of terror, Loki came trotting out of their shared cabin without a care in the world. Loki walked right up to James, climbed in his lap, circled twice, and flopped into what James assumed was Loki's favorite position on his lap. James smirked, then absent mindedly started stroking Loki's head. In his own head, he went over what he was going to do in the next the next couple of hours.

James had no idea how long he had been sitting there, but his wondering thoughts were interrupted by a soft whine emanating from his lap. When James looked down, he saw the two golden eyes staring back. "What's up, buddy? What do you need?" Loki's only response was another soft whine. "I bet you're hungry. Aren't you? I guess I could use a little food myself." James said while getting up. Half way to the food synthesizer, James stopped dead in his tracks, causing Loki to run into the back of his legs, "Did you just use the recycler? I only showed you how to use that once last night." James said while looking down at Loki. Loki looked up at James, sneezed because his nose just got jammed into the back of James' leg, and then barked once. He then proceeded to go around James toward the cockpit. James laughed as the pup walked right past the food synthesizer, thinking now he was leading the way. James punched in the puppy food recipe the vet had given him and waited 30 seconds for the kibble to fill the bowl. James left the bowl full of dog food on the counter and punched in a light breakfast for himself. Once he had both meals in hand, he sat Loki's food on the floor, sat down next to the little dog, and absently ate his food while looking at his deadly cargo.

After cleaning up, James could not think of anything else that he needed to do before he was going to have to act. With that settled, he headed to the cockpit and sat down in his chair. Not a second after his butt hit the padding of the chair, Loki was in it and making his little circles. James had to grab him before Loki plopped down. He held Loki up to eye level and said, "Sorry, buddy. You can't sit here. I do, however, have a little seat and clip for your collar in the co-pilot seat." James clipped Loki in and gave him just enough slack so Loki would be able to see out the plazsteel windows. Loki whined for a few seconds, then set his gaze out the windows.

The two-options James had to get to the fault were; to stay in atmo and take four hours to get there, or burn for space and shoot a reentry approach. The latter would only take about 90 minutes and save on his reaction mass. James chose the second option. Besides, there was something that he wanted to see.

The ride to the big black was uneventful, and soon James' computers told him that he achieved a stable orbit around Jin. James smiled as he slowly turned down the artificial gravity. He watched Loki intently to see his reaction. To his surprise, Loki stayed planted in his seat. James unstrapped himself and then reached over to unstrap Loki. Still, the dog did not float. James anchored himself with a foot under his seat and pulled Loki off. It took quite a bit more effort than James thought it would. When Loki popped free, the pair went tumbling through the cockpit and into the main corridor. By the time James recovered from the uncontrolled spin, Loki was standing on the ceiling, barking at him.

James laughed and called Loki to him to see how well the pup could move around. At first, Loki seemed to have trouble lifting his paws and moving around. However, within a couple of minutes, Loki was able to run and even began to jump from the ceiling to the walls and back again. James went as far as to throw a ball softly down the main corridor and watch Loki bounce from wall to wall to ceiling to floor, trying to intercept the ball as it floated lazily through the ship. While Loki did tumble a couple of times, he seemed to master the skill before it was time for the re-entry.

After an hour, the ship's computer reminded him that it was time to perform the re-entry maneuvers. James turned up the gravity slowly while trying to coax Loki from the ceiling. At .5 g, Loki could no longer maintain his grip on the ceiling and fell into James' arms. Loki started wiggling to the point that James had to let him go and quickly ran to their room. A few seconds later, the pup emerged from the room, and the waste recycler whirled to life. “Man, I am glad you did not do that while we were floating around,” James stated as Loki trotted past him and, jumped into the co-pilot seat and waited to be clipped in. “You‘re one smart dog.” Loki just sat waiting to be buckled, wagging just the tip of his tail.

As James started the re-entry sequence, a sense of dread started to wash over him. He rolled the spacecraft and looked up the see the planet below him. Even from space, he could see the Great Continental Divide. He pointed the nose of his ship to the most desolate place along the fault and descended through the atmosphere. The pit in his stomach grew with each mile he descended, but he had to push on. After all, this was his plan. This was his ultimate goal.

Chapter 8

Descending through the atmosphere, James was able to see a large portion of the hemisphere, from the oceans to the desert plains he intended to land on. As Jin drew closer, James begin to see large cities and their suburbs. Once he entered the clouds and everything disappeared into the white haze, James began to shake, thinking about what was about to happen. James sat in silence as his ship landed at his preprogramed destination. The gentle bump of landing startled James out of his thoughts.

As if James was on auto pilot, he moved through the ship into the cargo area. He pulled at Lucky Number Eleven only to realize that he could not move the XS-12 without some kind of mechanical assistance. James smirked as he realized he had never actually moved the containers in any type of gravity and grabbed a hover dolly from the side of the cargo area. James still had to figure out how to remove the container from the stack, he ended up rigging a pully system just to get the container onto the dolly.

After James loaded up the dolly, Loki woke up and wanted to be un-clipped from his seat. James dutifully set down the hover dolly and unclipped him so he could have the run of the ship. James then opened the ship's hatch to set up the core buster. As he did this, Loki bolted straight out and ran into the desert. James started to chase him, but stopped when he got out of the hatch. He saw Loki running around the ship and urinating on every scrub brush and rock he could find. James smiled as he watched Loki run. Once he turned around, all the happiness drained out of him. The XS-12 snatched his brief moment of joy. "If this is what I want, then how come every time I see the implement of my immortality, it fills me with dread?" He asked the wind. The only response he received was the faint sound of liquid hitting the dry ground.

James shook off the feeling and grabbed the dolly. He landed the ship about a quarter mile away from the wall of the fault line. His plan called for him to get the Core Buster as close to the ravine wall as possible. The heat was oppressive on his trip out there, and with every step, his anxiety grew. Sweat rolled off his head and body to the point where his jumpsuit was soaked, and now it was hard for him to see.

He only made it a few hundred yards before he was exhausted and done with the situation. He tilted the hover dolly and struggled to get the XS-12 onto its end, ensuring the arrow was pointing up. Without thinking, he pulled out the tablet, chose Option Two, set the timer for 90 minutes, and punched the 24-digit code. Just as he was about to hit the execute option, his finger stopped an inch above the tablet. James began to shake. He had almost condemned a whole planet because he was frustrated and hot.

James stood statue still, sweat pouring off his body, shaking uncontrollably, and on the verge of losing his breakfast. He just stared at the shaking tablet. Many thoughts running through his mind; all of the childhood neglect, the beating he took yesterday, the interaction with the receptionist at the veterinarian's office, being ogled by the store clerk, and finally, the kindness shown by Dr.

Robinson and Alyssa. He was so conflicted. The hurt of his first 50 years was contradicted by the kindness of the last two people he talked to. James just stood there on the verge of collapse, shaking and sweating in the desert heat, and then tears started flowing out of his eyes. They did not start as a trickle but a full-blown body retching sob. He could not do it. He could not bring himself to push the button. Too many lives would be lost. And for what? So that his name would be remembered?

"What was I thinking? What is wrong with me?!" James shouted to the gods.

It was as if a great weight was lifted from James' soul while the tears still flowed. The tears felt cleansing as they splashed onto the tablet. As James absent mindedly wiped the tears and sweat from the tablet. A 90-minute countdown timer replaced the 24-digit code and activate option. The timer strobed red every ten seconds with an abort button. The button was greyed out with 52 dashes under the icon. Panic. James panicked. Just as he decided to spare the planet his wrath, he accidentally condemned every person in Jin to a horrible death.

What was he going to do? He never got the abort codes. He could just fly off and leave the planet. Wasn't that the plan in the first place? His name would still go down in history, and no one would know it was an accident. James could claim that his plan went off without a problem and that he would do it again. The gods had cast the die, and this planet was doomed. That there was nothing he could do. The tablet beeped, and James looked at the flashing 89-minute timer. Panic again. Alyssa and her children's faces flashed before him in his mind's eye. Once again, he changed his mind.

James had to get a running start to knock the XS-12 over and onto the dolly. By this time, Loki was running around the dolly, thinking this was some kind of new game. James activated the hover dolly and began to run with it in tow toward his ship. Loki beat him there by a good margin and waited for him at the top of the ramp. While the XS-12 was now weightless for James, he still had to manage its mass and almost crushed himself between the five XS-12's in the cargo bay and the one he had in tow. At the last second, he shut off the dolly and jumped out of the way, the XS-12 crashed into the ground and then slammed into the other five XS-12's sitting in his cargo bay. James held his breath until he was sure the collision was not enough to set the munitions off.

The tablet sounded its minute beep. James looked down to see that there were only 85 minutes left before the Core Buster started its cycle of destruction. James pushed himself off the floor and started to run toward the cockpit, only to be tripped by Loki running underfoot. As James lay on the deck of his ship trying to catch his breath, Loki trotted up to his face and started licking his face as if to apologize for the fall. The red ten-second flash of the hardened tablet forced James back to his feet, and running once again to the cockpit. This time making sure to dodge the puppy darting around his feet. As James sat in the pilot chair, Loki jumped into his seat and waited to be buckled in. James initiated the start sequence and clipped Loki in. He set the engines to max burn and prepared himself for the G's the inertia dampeners could not compensate for.

"Sorry, Loki, this is going to get rough." The ship leapt into the sky at full burn, it was only seconds before the G's hit, and he started to black out. As the black tunnel closed around James' vision and his senses dulled, he felt the loose XS-12 more than heard it slam into the back bulkhead of the cargo area. His last thought was a fleeting hope that the impact did not set off the Core Buster.

James snapped back to consciousness with the sound of Loki's repeated barking. The second thing he noticed was that the G load was no longer pressing the air from his lungs. James scanned the panel in front of him and saw that his ship was still accelerating out of the solar system's elliptical plane, but the inertia dampeners had caught up. James grabbed the tablet and checked the timer: 22:39. He had been out for over an hour. James wondered what he was going to do now. He was almost certain that if the Core Buster went off, Jin would not be affected.

The thought of just continuing to ride out of the solar system and let the Core Buster explode in his cargo area had crossed James' mind. The thought lost all appeal once he looked to his co-pilot and saw Loki staring out the forward window. James had to figure out a way to get rid of the damn bomb. "Plasma torch! I can use the XS-12's plasma torch as a make shift rocket!" James yelled as he ran to the gear locker and started to don his EVA suit. Ten minutes later, he stood in the cargo hold and turned off the artificial gravity. When the artificial gravity turned off, Loki climbed the hatch separating the cargo area and the rest of the ship to see through the transparent plazsteel window. He watched James work franticly to untangle and move the XS-12 toward the outer hatch.

Once at the hatch, James tried to open it. Instead, he was met with an error code on the door panel. He had forgotten to pull all the air out of the cargo area. Quickly, he started the pumps. In a matter of minutes, the cargo area was a vacuum. James looked at the timer on the tablet: 4:27. This was it. James almost went out the door without clipping in. He caught himself before he made that mistake. James attached a line to the XS-12 and stepped out the door.

As he cleared the inertia dampener field, he was slammed into the side of the ship due to the continued thrust of the engines. James almost blacked out for the second time that day. With the XS-12 pinned right next to him, James brought both arms together and triggered the remote function for his ship on his left forearm. His built-in display pulled up several options, one of which was shutting the engine down. James was just able to shut them down before his vision started to tunnel closed. As soon as he killed the engines, James was no longer pressed into the side of the ship, and his vision returned to normal. He checked the synchronized timer on his HUD. The reflection on his visor told him that he had less than two minutes before the Core Buster started its drilling sequence, and James certainly did not want to be attached to the munition during that phase of its deployment.

James took a deep breath and pulled the line that connected him to the Core Buster, he forgot that he could not just pull the XS-12 toward him without creating his own movement. The Core buster crashed into him with a much greater force than he expected, and he slammed into the hull of the ship as the mass of the XS-12 knocked the wind out of his lungs. James just learned that he hated space and all of its made-up rules. Coughing and nearly throwing up in his helmet, James forced himself to take deep breaths to regain control of his own body. The one-minute timer started blinking on his HUD. James had to act fast. He oriented the XS-12 with the *THIS END UP* end pointed toward the solar plane. He then started to spin the Core Buster as fast as he could along its long axis. The timer hit zero.

The bottom of the container began to glow a violent yellow and red just before a bright blue plasma jet erupted from the end of the green container. The XS-12 slowly unscrewed itself from its container and stopped exiting the container once the grab wheels could not gain pursuit on the ground it was designed be placed on. The blue plasma jet continued burn as the whole weapon slowly gained speed away from any form of mass. This was about as good as James could ask for and he pulled himself inside his ship. As he collapsed and floated just above floor, he brought up his remote and set his ship to the maximum safe burn. He would be damned if he was going to black out again.

With the inertia dampeners working, James did not even feel his ship change directions when he engaged the main engines. He re-pressurized the cargo area, removed his helmet, and opened the internal cargo bay door. Loki jumped off the door, bounded off a wall, and then collided with James, sending them both in to a slow spin. James was too tired to care. He just held Loki and gently spun in the cargo bay.

Chapter 9

An insistent beeping pulled James from his floating post-adrenalin crash. His whole body hurt, and his mind was in a thick fog. It took all the mental power he had remaining to search for the source of his annoyance. James held up his left arm and saw another timer counting down. He pooled all his mental faculties and discovered the reason for the timer. It was his best guess as to when the XS-12 was going to explode. The manual said that the XS-12 could take up to 90 minutes to reach the core of a planet. Since the Core Buster was in a vacuum, he had no idea when it would actually go off. The timer currently said he had 72 minutes left before there would be an explosion of unknown strength.

James made his way to the cockpit and locked the rear tracking system onto the Core Buster. There was only 17 minutes left on the timer when James felt a lurch in the ship and was almost thrown into the console in front of him. His speed was dropping as the singularity pulled him deeper into its gravity well. Loki was thrown against his restraint and let out a yelp as the harness pulled against his neck and chest. Loki looked around to find the force that threw him against his restraint and seemed disappointed when he could not find anything. A moment later, the monitor showed the stars behind him bending. The radiation sensor spiked. A second later it felt as if the ship was fired out of the cannon. The last thing James saw behind the ship was what looked like a star going super nova. Then, the inertia dampeners overloaded, and he blacked out for the second time that day.

As James came to, he mumbled, "I am getting really tired of waking up when I didn't choose to go to sleep." He looked over to Loki. The little dog looked back, not seeming to have a care in the world. James switched his attention to his consul and started running diagnostics on all the ship systems. This would take about an hour before the computers returned the results. James unclipped Loki and started to head back to his gear locker. Loki bounded off the ceiling and walls, making sure to cross James' path as close as he could. At one point, Loki used James as a spring board to change directions and pushed James off his course into a bulkhead. James just laughed, "At least someone likes space. I'm pretty sure the universe is trying to kill me." James stripped out of his EVA suit and hung it reverently in its gear locker. "When we hit a port or space station, I am going to find a way to get you cleaned and upgraded," James said to now one of his prized possessions.

Then it struck him. James realized he had no plan, no money, and nowhere to go. James was numb and just floated there, slowly drifting inside his ship. He knew exactly where he was in space, but he had no idea what was going to happen next. For the next month and some change, James spent his time working out, playing with and training Loki while avoiding any type of communication. As he drifted through space with all of his external systems powered down, the little recon ship was all but invisible to anything but a direct scan.

James and Loki worked through the training videos that Dr. Robinson had given them in what had to be record time. James swore that Loki could watch the vid and execute the skill without even practicing. By the time they worked through the end of the vid's, Loki could have passed as the universe's best trained service dog. James almost got to the point where he was going to test the doctors claim that Loki could survive in a complete vacuum. He went as far as to put on his EVA suit on and enclose both himself and Loki in the cargo area. Just as he was about to pull out all of the air out of the cargo area, he exclaimed, "What the hell is wrong with me? This could kill you. I think... I think we have been out here long enough, bud." James removed his suit and huffed as he walked back to the pilot seat and flipped on the comm unit.

The screen lit up with hundreds of messages. The shear amount of news feeds and junk messages James received was overwhelming. The only personal message he received was from Dr. Robinson asking how Loki was adjusting to his new life. After James read Dr. Robinson's message, he started applying filters to the rest of the messages in order to find a place to begin the search to see if anyone suspected him. After a few minutes, he just decided to watching a couple of news feeds. The most popular theories for the explosion ranged from an exotic new weapon malfunction to the start of an alien invasion; some religious sects even proclaimed it was an act of God.

None of the news feeds even hinted at the ancient weapon being the cause of the temporary star and the subsequent explosion; and absolutely no mention of him. James was conflicted. He still wanted the universe to know his name, but he knew that he did not have the skills, contacts, and, most importantly, the money to run from the law. So, he was relieved, as well. James turned to Loki and asked, "Well, what do we do now?" Loki just stared with his head cocked, alternating sides from time to time. "Why does it feel like you know the answer and are just messing with me?" James asked, halfway expecting Loki to answer. "Well, I guess it's back to Jin," James said, feeling completely lost, "at least we know the area. And I doubt we will run into Hunter."

James orbited Jin, still trying to decide whether or not to land on the planet or just dock with the station. Out of the corner of his eye, James saw his comm flash to indicate that he had received a personal message. Happy to have the distraction, James hurriedly opened up the message.

Was it you?

This was all the text read. It was the sender that made James nervous. The man only went by "Ski". He was the arms dealer that sold James the XS-12's, and now he wanted an answer. James no longer worried about running from the law. He had to worry about what he was going to do with Ski.

After much deliberation, James decided that there was no point in trying to lie or run. Ski was too well connected and had informants everywhere. James typed a single word reply, *Yes.* His finger hovered over the send icon and laughed, "Last time I hesitated like this, I almost destroyed a planet. Well, here goes nothing." James pressed the send icon, leaned back in his chair, and patted his lap for Loki to hop on. James grunted as Loki landed on his lap. "Buddy, I didn't realize how big you have gotten. You have got to be 50 or 60 pounds by now. You may not be a lap dog anymore." Loki paid no attention, made two awkward circles, and spilled over James' lap as he laid down.

James decided that the best place for him to be right now was the orbital station around Jin. The docking fee would use almost the rest of his money, but it was ultimately cheaper than burning the fuel to perform a re-entry and another take off from the planet. Three hours later, James was docked on one of the cheapest berths on the outer ring of Jin station. James set all of his ship's systems to standby when his comm notified him that he had received 50,000 credits into his account. With that much money, James could fuel up and supply his ship for more than a year if he was careful with the credits. Less than a minute later, James received a second message from Ski. It was just coordinates and a date three weeks from that day.

James booted up his nav computer and input the coordinates. The computer came back with a location in the center of an asteroid field a few solar systems over. It would take about a week to get there. That gave James the time and money to figure out his next move. James knew that whatever he did next, he would have to be very careful with his next move. Ski was a very dangerous man.

Chapter 10

True to his word and flush with cash, James grabbed his EVA suit, put Loki on a leash, and headed to the nearest suit repair shop on the station. As James navigated the station, he was stopped every few minutes by someone who wanted to pet or play with Loki. At first, Loki was shy and just tucked in behind James' leg. By the fourth or fifth person, Loki would just look at James to make sure it was okay and then do a half prance half hop over to the person that wanted to pet him.

After much longer than James wanted, they arrived at the suit and robotics repair shop. Upon entering the shop, the pair was greeted by the smell of well used garage. Parts for suits and all manner of mechanical things were haphazardly displayed in stacked crates that created aisles. A rotund man sat behind the counter, soldering a part of something, its use completely foreign to James. As James approached the counter, the man said without looking up, "It will be 3000 for the cheap suit, 5000 for my best work."

"What? Wait, I already have a suit. I just wanted it upgraded and routine maintenance done," James stated.

"That's boring, and you already have the top of the line suit. I can clean it and replace all of the seals and filters, but that is about all. That will be 200 credits, and I can have it back to you in a week. 300 credits will get it to you tomorrow." The man said, losing interest in the routineness of the conversation.

"Well, what was with the 3000 and 5000 credit quote when I walked in?" James asked.

"The quote was for me to build the dog a suit," the man said, starting to show more interest again.

"I never thought about getting him a suit. To be honest, I did not even know that it was possible. What's the difference in the 3000 and 5000 credit suit?" James asked.

"The 3000 credit suit will be a basic gray suit with a tether attachment. You will be able to monitor all life support systems, and it will have a short range comm so he can hear you. The 5000 credit suit will, um... Well, hell, what do you want it to look like? And what do you want it to do?" The man gave a slight pause and then began again, "I will caution you, though. If you did not spring for the advanced intelligence package for your dog, the 5000 credit suit may be a waste."

"You've done this before, right?" James asked, now very interested in what the man had to say.

"I thought that is why you came here. I don't mean to brag, but I am one of the best custom suit builders within this cluster. But that is a moot point, unless... what's the dog's name?" The man asked.

"Loki. By the way, my name is James... Aaaaaand you are?" James inquired.

"I go by Port. Anyway, I have built a ton of dog suits. But I really need to know, how old and how smart is Loki?"

"I was told by the vet that he was a very smart dog. As for the age, I really didn't ask the vet when I brought him in." James answered.

"Okay, when did you order him and pick him up from the vet?" Port asked, starting to get annoyed with the way James did not know the basics about his own dog.

"Look, I found Loki on the street. He was in bad shape, so I took him to the vet. That was over a month ago." James stated.

"Oh, okay. That explains some of the gaps in your knowledge. Here's what I am going to do for ya. I'll take Loki into the back and run him through some tests that assess his age and intelligence level. But, I will only do that if you pay me 1000 credits or sign a contract that says if Loki is smart enough to use the custom suit, you will buy one," the now *very* interested Port said.

"What happens if he is not smart enough?" As James asked the question, Loki snapped his head up, stared at James, then gave a little huff.

"Well, judging by Loki's reaction to that question, I don't think that we are really going to have to worry about that," Port chuckled.

"The vet said he was going to be very smart. I guess it's time to test that claim. How long is this going to take?" Asked James.

"The test takes about an hour. To build the suit, a full week. And there is no higher price for a faster delivery time, that is as fast as I can do it." Port said.

"What about my suit?" James asked, remembering the original reason for going there.

"Uh, oh ya, 100 credits on top of the custom suit price. So, do we have a deal or what?" Port asked.

"Where do I sign?" James asked happily. Port held out a tablet, and James placed his hand under the scanner. After a second, the tablet beeped. Port looked at it and smiled. Port struggled to his feet and said, "Alright, I'll take Loki, and you can go do something else for an hour or so. I'll call you when we're done."

"Can I just go back there with you?" James asked.

"No, it could affect the cognitive tests. You'll just have to go somewhere else. There is a bar and a restaurant within walking distance of here. Or, if you are looking for some more intimate company, there is a VR facility and a flesh palace a short ride on the tram from here," Port said over his shoulder as he took Loki's leash and led him into the back of the store. James suddenly felt very lonely. This was the first time that he and Loki had been separated since James picked him up from the vet.

James passed the time by checking on his ship's refuel and refit status. After seeing no issues and having 58 minutes left, he wandered aimlessly, looking for something to do. In the end, he decided go to the bar and order some food. The bar was a lot cleaner than James expected. The bartender took his drink and food order with a friendly smile and polite conversation. James mostly just moved his food around on his plate and continually checked his comm. James finished his meal and paid, the whole time, his comm stayed never left his sight.

James had almost given up on the message coming through, thinking that somehow Port had lost his comm address or maybe something happened, and he needed to get back there. James did not have to go far. As he rounded the corner to Port's store, he saw Loki through the plazsteel. Loki had not seen him, so James stopped and watched for a few minutes. It looked like Loki was wearing an odd looking collar and was intently watching something James could not see. Suddenly, Loki bolted from sight, and James heard a series of muffled barks in what seemed like a repeating pattern.

A few moments later, James' comm buzzed made him jump when it interrupted his pondering on what Port could be doing with Loki. James shook it off and looked at his comm. It was Port, and the screen simply said. "Stop standing out there and come in. The tests are over." James entered the store, and Loki was right there, wagging his tail so hard that the tip of his tail almost smacked his own face. "Well, looks like you owe me 5100 credits," Port said through a big smile.

Over the next week, Port ran James and Loki through drills and mock fittings to make sure that the pair knew their gear and emergency procedures. Port added a software update to James suit that allowed him to control every function of Loki's suit. Loki's suit had a built-in comm that was a two-way encrypted channel radio. This would allow James to talk to Loki. Loki could respond with one of three buttons that was built into his helmet that he could press with his nose. One button to Loki's left, one below his chin, and one to his right. James had Port program the buttons to activate a red, yellow, or green icon that would flash on

James' arm comm when corresponding button was pressed. James worked out a simple system of commands and questions that Loki would understand and could answer by pressing either the red button, meaning *No*; the yellow button, meaning *Maybe/I don't know*, or the green button, meaning *Yes*. With this simple communication system, James began to understand just how smart Loki was. James liked buttons so much that he paid Port to make a collar that functioned in the same manner as the helmet switches. James had both the collar and helmet linked to his comm, his EVA suit, and a small embedded light ring in the collar.

Toward the end of the week, James asked if he could commission Port to complete one more job. However, this one would be a bit odd. At first, Port was worried about what the weird project was. After James explained it, Port agreed to build them for only a few hundred credits each. Now, James had the beginning of a plan.

At the end of the week, James was so excited to get both his and Loki's suit back that after the final fitting, he kept everything on except the helmets. While he was hardly the only person wearing an EVA suit in the station, he was the only person wearing matching black and orange EVA suits with a dog. James and Loki almost strutted the entire way to their ship.

The fact that James and Loki became a common sight on this route, and that they were now wearing matching EVA suits, only added to the amount of people that wanted to come play and pet Loki. Their final trip to the ship took twice as long compared to their normal time, and James could have not have been happier. He finally felt like he belonged somewhere. James did not even care that most of the people got to know him because of Loki, at least they got to know him.

Chapter 11

James left the station over Jin and headed to the nearest que-drive exit point. The exit point had to be out of any gravity well larger than an asteroid. Otherwise, the que-drive would not be able to form the warp bubble to enable the ship to travel faster than the speed of light. It would take James six hours at a moderate burn rate to get to the jump point. After reaching the jump point without incident, James plotted a course that would take him to the opposite side of the system that Ski had indicated. As soon as the warp bubble formed around the ship, James got Loki into his EVA suit and did the same for himself.

James pumped all the air out of the ship and into the holding tanks. He then turned down the gravity to zero. With his cargo area packed full, space was limited. However, James managed to put a harness over Loki's suit and attach it to one of the XS-12's. Loki and James worked through the entire ship with Loki, pulling and maneuvering the XS-12 into different locations, then detaching and making his way to a marked area in the ship. Every day, James added more complexities to the movements, and by the end of the week, Loki could get in the harness without assistance and move the Core busters anywhere around the ship without banging the munition around. Then, Loki would disconnect and make his way to wherever James indicated. James' plan was coming together. He just hoped that it would work.

James' ship flashed into the solar system indicated in Ski's text, and just like James planned, he was on the opposite side of the asteroid field and week early. James turned all his passive systems onto the location of the coordinates that Ski gave him. There he discovered what looked like an abandoned ore processing facility nestled in a tight group of asteroids.

"Ok, buddy. I think I see the place where we're going." Loki, not impressed, simply circled in the copilot seat and then laid down. James turned off all external systems, and once again, the ship almost disappeared as it drifted through space toward the processing plant. A day after arriving, James and Loki suited up and set almost half of their cargo on a small asteroid when they were blocked from the view of the mining facility. As quickly as they could, they got back into their ship and resumed their drift toward the plant.

During the next week as the ship drifted closer to the plant, James could see that the plant was not abandoned. It was just made to appear that way. James observed ships coming and going from the facility, but they always took the same route. James began to wonder if that was the safest route or if there was defensive perimeter like he saw in some of the vids he enjoyed. Just to be on the safe side, he decided that it was best to assume there was some sort of perimeter and he would have to watch his step very closely. On the day that James was expected to arrive, he landed on one of the asteroids close to the coordinates he was given. He turned on all his active systems to announce his presence to the system. Now much closer, James knew he was being watched. So, he turned to Loki who was in his

EVA suit and said, "Well Loki, I really hope you are ready for this." A small green light illuminated on Loki's collar when he triggered the embedded switch.

After a solid day of just sitting there, James' ship notified him that there was a tight beam laser comm link request. James accepted the request and was not surprised to see Ski's ice blue eyes staring at him through the screen. "Well, good morning, James, you gave us a bit of a surprise, just popping up like that," Ski's deep voice hinted at the anger in his eyes. "You know, I am not very fond of surprises. But let me take a guess on how you pulled that off. I know you left the Jin system two weeks ago with... What was his name?... Oh yes... Loki, I believe," Ski continued on, "You must have gotten here around a week ago to be beyond my sensor net, then you would have to drift through the asteroid belt and use cold jets to land on that asteroid. How did I do, James? Am I right?" James' blood ran cold. He knew that Ski had contacts everywhere, but to know his dog's name and just how James had gotten that close to the truth made James shudder. James painted on a smile, "That's not too bad. Just off with a few minor details, but close enough."

The camera panned to follow Ski as he walked around what looked like some sort of control room. Ski stopped behind a woman in a jumpsuit seated at a terminal typing commands in. He casually pulled a small laser pistol from his belt and shot the woman in the back of the head. She slumped forward, and her head hit the keyboard in front of her. Blood seeped from a small smoking hole just below her right eye. "You see, James," Ski said as if nothing had happened, "She was in charge of monitoring the security of this facility. Apparently, she was not doing a very good job." James stared in shock. Did that just happen? James knew then that he was in *way* over his head, but his death was cast, and now he had no choice but to push ahead with his plan.

James mustered all the bravado that he could. "I am glad I helped uncover a gap in your system, so you're welcome. If I had any ill intent, I would have been on your doorstep before you could have stopped me." James said as if he did not just witness a murder.

"It's cute that you think that, James. It gives me hope that I won't have to kill you, too." Ski said, returning to his seat.

"Look, you called me here. Care to explain what you want?" James said as forcefully as he could.

"Be careful, James. Remember, I hold all of the cards here." Ski commented.

"That's odd because the way I see it, I have 12 of the most powerful weapons ever invented at my disposal. And that, my friend, seems to be one hell of a trump card." James proclaimed with convection in his voice.

"I didn't know you could grow a spine in such a short time, James. I don't like it. But since you're right and you do hold the... What was it?... The *trump* card was it? Anyway, last I checked, I only sold you six Core Busters and you just said 12 of them. That's interesting." Ski paused, and James took the opportunity to speak.

"What? Do you think you were the only one that claimed to have them? Turns out there were a couple of others that actually had the Core busters. And to tell the truth, yours were on the cheap side of what I paid for some of the other ones."

"James, I *do not* like it when people lie to me!" Ski shouted. It took everything in James' soul not to buckle under. Instead, he nonchalantly said, "Believe me or not; makes no difference to me. Now, what do you want? I have other places to be. So make this quick, you've wasted enough of my time." James could see Ski beginning to shake with rage, and just as suddenly as it appeared, the rage vanished and was replaced with an icy veneer.

"Very well, James. I want my Core Busters back. And you are going to give them to me." Ski said through gritted teeth.

"I am not sure you understand how *selling things* works." James' voice almost cracked and he had to make himself stop shaking.

"I'm done talking. Transfer the weapons over to me, or I will destroy your ship and be done with this!" Ski's veneer was gone.

James brought his tablet into view with his thumb on an icon. "Do you recognize this? Fun thing about this is the deadman switch. Just in case you forgot, let me remind you. If my thumb leaves this icon for any reason, one of the Core Busters will turn this place into a supernova. So, tell me again what you wanted."

James noticed the blood drain from Ski's face as he stood there in shock. "Ok, James, maybe I'm not done talking. I'd like to buy back the weapons I sold to you, and I'll pay half of what I sold them to you for. I think that sounds pretty fair, what do ya say?"

"I say no." James paused, and just as Ski opened his mouth to speak, James started again. "You sold me an untested weapon that you didn't even think would work. I took the risk and proved the weapon was a real threat, making them far more valuable. So, the deal I will make you is for the full price of what I paid for six of them. I will sell you one Core Buster with the option to buy more at a later date at the same price. And please remember, if I so much as sneeze, we all stop existing."

James could see Ski grinding his teeth in frustration, "James, be reasonable. I cannot pay that price for a single Core Buster. I-" Ski started, and James interrupted, "Then I guess we are done here. Have a good day."

"STOP INTERRUPTING ME, "Ski roared. James sat back, smirking outwardly, but inwardly, he was praying that he did not overplay his hand. "Well?" James asked.

"Sell me two of them for the full price you paid, and I will buy more at the full asking price at a later date. I cannot afford to pay full price for just one. I need at least two to be able to make enough money to afford more." Ski said, trying to regain control of the conversation.

"I can live with that. I'll wait here until you can get me the creds, then you can come over here alone to get your Core Busters." James killed the connection and sprinted to the cargo area, where Loki was suited up and waiting in his harness. James knew this next part was going to be very risky but there was no turning back now. He secured his helmet and double-checked Loki's suit, then waited until the asteroid rotated away from the mining facility.

Once he was no longer visible to the ore processing plan, he opened the hatch and bounded out the door with Loki on his heels and XS-12 floating on its tether. Once on the surface, James noticed the slightest bit of gravity, but not enough to bring him back to the asteroids surface any time soon. James began to panic as the ground got further and further away. As James looked between his booted feet, he saw Loki running along the ground without coming off the asteroid's surface more than a few feet between bounds. Once again, James cursed his lack of forethought. How did Loki know to turn on maneuvering nozzles to stay close to the ground while he was flying around 20 feet off the asteroid like an idiot?

James took a deep breath and pulsed his maneuvering nozzles to get back to the ground at a reasonable speed. By the time James found his rhythm between bounding and pulsing his cold nozzles, Loki had already reached the area highlighted on his Heads Up Display (HUD) by James. He watched as Loki jumped to the side and turned to face James. The XS-12 coasted by Loki and slowly started to descend before the tether went tight and jerked Loki backwards. James laughed out loud as Loki and the XS-12 touched the surface of the asteroid. "Loki, are you okay, buddy?" James' HUD pulsed green one time to indicate Loki was okay.

"Now it's my turn to do some work, but I have to hurry the mining facility will be coming over the horizon soon." Loki had already disconnected from the XS-12 and the cable automatically retracted into his suit. James quickly gathered all the loose rocks around him and began piling them up around the Core Buster. Soon, all James could see was the words on the end cap of the munition: *THIS END UP*. "Time to go back, I'll race you." By the time the green light in James's HUD blinked, Loki was moving faster than James had ever seen him move before. "Loki, be careful. Don't go too fast. I don't need you getting hurt." All James got in response was a single red light blink in his HUD as he tried and *failed* to catch up.

Loki was waiting by the ship when James finally bounded up to the ship. "You cheated," James gasped out. A single red light was Loki's immediate response. "Fine," James said as he opened the hatch to the ship. James and Loki moved the remaining four XS-12's out of the cargo area and into the ship's main corridor. With that task complete, James located the emergency escape hatch on the floor and pulled the release handle. The hatch swung in, revealing the rocky dust covered ground of the asteroid. "This is going to be a huge pain and an even bigger mess." James grabbed his pick axe and started in. After four hours of back breaking labor, the hole was deep enough for Loki to position one of the remaining XS-12's into it. James quickly covered the hole and packed the dirt down as tight as possible. By the end of the evolution, James still had a large amount of dirt and rocks in his ship. In the end, he had to clear out a small maintenance space and put all the left-over regolith in there.

After thoroughly cleaning the cargo area, James and Loki moved the three XS-12's back to their original resting place over the emergency escape hatch. James turned the gravity to 1 G and pumped the air back into the ship. James and Loki exited their EVA suits, and James hung them in their gear locker. "Alright, buddy. We should have a solid day before the funds clear the banks. Let's get some food and rest. Tomorrow is going to be another long day." A green light illuminated on Loki's collar as he walked toward the small dining space.

Chapter 12

James had a fitful night of sleep, mostly tossing and turning instead of sleeping, even with Loki trying to calm his nerves. There were just too many things that could go wrong with James' plan. Too many ways for Ski to see through the deceit. After a few hours, James gave up and went over his plan and tried to patch all the holes he found in it. Without being on a planet where there was no real day and night cycle James could use to tell time, James instead watched the star at the center of this solar system cross the sky every 5 hours, but it never fully went below the horizon. He knew that to transfer that many credits it could take several days. After all, he had completed that same transfer just over two months ago.

James took the next couple of days, ensuring that Ski was not trying to double cross him. The delay for space-net was about an hour, so James could access all the intergalactic sites, but none of the sites were in real time. James set up an alert on his banking account to notify him of deposits. When that happened, he set an automatic transfer of the funds into several different accounts that no one else had access to. With that done, James scanned all of his systems for the fifth time, making sure that all his active sensors were set to full and nothing was approaching.

Two message notifications hit James's screens at the same time. The first was a notification that the credit transfer was complete and was now being dispersed across all of his accounts. The second was from Ski. James let the tight beam through. "You have your credits, now I want what's mine," Ski spoke without waiting for James to speak, "I am coming over with a team, and you will give us the Core Busters."

"There will be no team and no weapons. Just you. Please don't argue, remember, I still have my trump card." James coolly answered.

"Okay James, I see you have a plan on how you want this to go. So how about filling me in so we can just get this done."

"No problem," James spoke with more confidence than he felt, "I saw a small shuttle with the hull identifier APS-16354 entering your facility about four hours ago. You are to board that shuttle within the next hour and land half a mile from my ship. Land your shuttle 90 degrees to the right of the nose of my shuttle. There is a small clearing there and I will be able to see you the entire trip over. I will meet you outside and transfer the two Core Busters over to you without you setting foot on my ship. Are there any questions?"

"I will not be able to carry both of the Core busters, even in that light gravity. I need to bring a grav-sled."

"I'll tell you what, Ski. I will throw my hover dolly in the deal for free. After all, I am a very wealthy man again." James smiled.

"Fine. See you in an hour, James. I look forward to seeing you in person."

"See you then, Ski."

James and Loki suited up and then loaded two XS-12's and their tablets onto the hover dolly. James pulled the air out of the entire ship and turned off the gravity. James now weighed about two pounds and had to be very careful not to accidentally crash into the ceiling. Once outside, James watched the small shuttle land right where he had directed. Ski wore a massive black EVA suit. It was so large and well armored it almost looked like a Mech. As Ski approached, James could see the hard point mounts for the weapons. With the scuffs on the mounts, it looked like the weapons were removed in a hurry and recently. James knew this had to be a Mech, but it was the smallest one he had ever seen. James knew what his next purchase was going to be. He wanted that suit. The only problem was the only person that he knew he could get it from was Ski. James did not think that Ski was in a selling mood. Well, at least not to him.

"Where are my Core Busters?" Ski demanded. James side stepped and answered, "Right behind me. By the way, what are the chances I can get one of those fancy mech suits?" Ski was an imposing figure out of the suit, standing a full head taller than James. In his EVA suit, James felt woefully outclassed. "I have to see if I have any in stock if you survive long enough to buy one." James could feel the hatred rolling over the radio waves. "Ski, I want us both to walk away from this deal. We can both be rich and powerful beyond measure, all we have to do is stick to the agreement."

"If that was the case, why did you bury a Core Buster on the other side of your ship? It seems to me that you are planning to double cross me."

"Well, I was hoping that you wouldn't find that one yet. But since you did, I will come clean. It's my insurance policy. I was going to leave that here and then drop off the tablet for it at a different location." James held up the tablet with his finger on the deadman icon.

"So, we were going to just give me a third core buster for free, huh?" Ski asked.

"Well, maybe not for free, but at a greatly reduced rate. After all, you would have the Core Buster, and I'd have the tablet. Both are worthless by themselves," James continued, "But since you found me out, I think all of my cards are on the table." A red light blinked four times on James' HUD. James quickly pulled up the video feed from Loki's view and saw four men taking cover behind the boulder strewn area on the opposite side of James' ship where Loki was standing guard.

"I see that I'm not the only one breaking the rules. Four of your guys are currently approaching my ship from behind me, and it looks like they are all armed. So much for the rules. What do we do now?" James asked as he held up the tablet. "I suggest your men back off. We can both win, or we both can lose. If either one of us slips up, this entire asteroid belt will be unrecognizable."

James saw Ski's square jaw moving as he spoke to the men approaching his ship. The feed coming from Loki showed the men headed back toward the XS-12 that James had partially buried. "I have called off my men. Let's get this done, I'm tired of playing these games. Give me the tablets and the Core Busters, and I will be on my way. James handed over the tablets, and Ski looked at them closely. "Why are they labeled 11 and 3?"

"That way, I could make sure that I didn't mix them up, see," James pointed to the XS-12's next to him, showing Ski the corresponding numbers painted on the sides.

"Are the abort codes still on the back of the tablet under the case?" Ski asked as he peeled the rubber padding off the tablet. James was stunned. He had the abort codes this whole time, Ski had simply omitted telling him. Once again, all the deities, past, present, and future, felt the full wrath of James mental tongue lashing. Ski laughed as he showed James the back of the tablet with the 52-digit code printed on a label, plain as day. "Did I forget to mention that? I'm sorry, I hope you can forgive me." Ski laughed as he replaced the protective shell on the tablet, laughing at James.

"Yeah, that must have slipped your mind." James growled out while trying to get a hold of his anger.

Ski peeled the back off the tablet labeled 3 and looked at the code, "This was one of mine also. James, did you really get other Core busters?" James quickly replied, "I am selling you what you sold me, I hope the rest of the Core Busters you sold me are as real as the first one."

"Fine, I don't really care one way or the other. We're done here." Ski barked as he grabbed the two XS-12's off the hover dolly with one arm and started back toward his shuttle.

James called over the radio, "Just remember. My finger is on the button. Remind your guys hiding in the rocks of that fact. We both lose if something happens to me." Ski made no indication that he had heard James' transmission.

Now, all James had to worry about was Port's code. He hoped it was as good as Port claimed. James looked at Loki's video feed and saw the four heavily armed men moving toward the fake XS-12 James left for Ski. "Let's go back to the ship Loki." As James took off from the asteroid, all he could think about was Port's code. "It had better work, or I'm in big trouble." A yellow light blinked on James' comm. "Sorry buddy, I was talking to myself." Now, a green light lit up on James' comm.

Chapter 13

James knew he was on borrowed time, and the only thing stopping Ski from capturing, torturing, and killing him was the fact that the XS-12 came with a deadman switch. James had to disappear and do it quickly. With his money back, he could easily do it. The only real problem he had was his family. Although James had no love for his family, he did not want to see them used as leverage against him. Luckily, James had made lots of shady contacts in his hunt for the XS-12's. He just had to find ones not on Ski's payroll.

On his way out of the system, James picked up the cargo that he had left over a week ago. All of the real XS-12's where right where he had left them. Now, he had to decide where to go to begin the planning the next part of his life. James immediately ruled out Jin station. *After all,* he thought, *Ski knew exactly when I left, what I did there, and who I talked to.*

After going through all of the solar systems James had dealings with, he decided on Hel. It was a fringe system that was an old supply depo for explorers going past the known galaxy. It would take James a month at his ship's max speed in a warp bubble. James had just enough fuel and consumables to get there.

On a whim, James sent his family a message saying that he was going to explore the universe and that he was not sure when or if he would be back. The message also contained an account number for each family member that contained 100,000 credits. Compared to the money he got back from selling the fake XS-12's, this money was insignificant. The message and funds could not be transferred while James was traveling faster than the speed of light. They waited in his queue until he was out of the warp bubble and in range of a relay station. Before each jump, James' ship would take a snap shot of the selected portions of the space-net. During his first two weeks traveling in the bubble, James found several small companies that he was going to purchase. After that, he searched for the largest single man long range craft he could find.

James' search went so well that he dropped out of the warp bubble early and visited a nearby station with a less than stellar reputation. K-19 was an old colony ship that dated back to before the que-drive was invented. The legend was that after traveling for two hundred years, the K-19 finally reached its destination only to find an established human colony on the planet. The planet was inhabited by colonists who came by que-drive, taking the trip in less than two years. The massive colony ship was all but abandoned and left in a high orbit. Many years later, it was bought and fitted with a que-drive to serve as an outpost for a mining super conglomerate. After it failed to turn a profit, it was once again all but abandoned. Slowly, it began to attract less than reputable inhabits and became a pit stop for explorers and pirates alike.

Nowadays, K-19 is run by three groups feuding over the right to claim the station as their own. Outright war was not common between the factions, but at least once a week, someone died or disappeared in a suspicious way. James had a couple of dealings with this place when he was initially hunting down the XS-12's, and none of the contacts here had any connection with Ski that James could find. When James dropped out of the warp bubble, his comm notified him that all of his messages were sent and his incoming messages updated.

The only message that concerned him was the one from Ski, wanting to know when and where he could pick up the tablet that was paired with Core Buster marked 9. James laughed to himself, Ski already had the tablet for Core Buster 9. James had just relabeled it as Core Buster 3. The real XS-12 was buried on the asteroid where James had landed. The first time Ski tried to blow up a planet with any Core Buster's, his center of operations would be pulled into a singularity and then blasted across the solar system.

In order for his plan to work, James had Port make him six Fake Core Busters and tablets. Then label them: 1, 2, 3, 4, 5, 9. The tricky part was getting Port to code all of the fake tablets to interact with the real tablet 9. James explained that the munition was for destroying stray asteroids and that the decoys were for a trap to lure out a couple of warlords. At first, James thought Port believed him. However, as James looked back on the situation, Port knew he was lying and said nothing. He just wanted the credits. On a whim, James grabbed one of the fake Core Buster tablets and pulled off the outer shell. To James' great dismay, there was a fake 52 digit labeled on the back. "You've got to be kidding me? Am I the only one in the universe that does not check under the case? Well, I guess the fakes were worth every credit." James laughed at himself, Loki just watched James for a few seconds, then laid his head back down.

James chose to ignore the messages for the time being. He had other things to get in order before he dealt with Ski. Instead, he requested a berthing for his ship on the most expensive dock on the base. He needed to make an entrance, but as his small ship pulled into its berth, he realized he may not have drawn the attention he was looking for. A small army of vendors and maintenance workers watched as one of the smallest ships at the station docked at its largest berth. James' message box was flooded with all manner of requests ranging from marriage to the deepest, darkest pits of human imagination. James remembered why he ended up hating most of the trappings of money.

After filtering his messages down, James picked out the few that interested him and sent out requests to be picked up at his ship and taken to the shops. James and Loki suited up, minus their Helmets, and headed out of the ship. James ensured that not only was his ship was locked up, but it was guarded by a local group of armed men called the Keller group, who had a reputation for performing their jobs without asking too many questions. As James walked past the group of guards, one of the men stepped in front of him and said, "If you come back with a guest, I will need you to let me know if you are being held hostage or if they are really your guest. When one of my men steps in front of you and asks who your guest is if your response is 'This is my dear friend,' that will let us know that you are a hostage, and the person or persons with you will be shot on site," the man continued, "all other responses will be ignored and you and your guest will be allowed to pass, do you understand?"

"That seems a bit extreme, doesn't it?" James asked.

"Mr. Minder, that is what you paid for. We could incapacitate them if that would make you feel better. But rest assured, the people on this station may not afford you the same kindness." The man said deadpan.

"I would rather not kill someone by accident if it can be avoided." James almost burst out laughing. Just over a month ago, he almost destroyed an entire planet so people would remember his name. Now, here he was, asking for guards not to kill his potential kidnappers.

The guard stared at James for a few seconds, "Fine. But if any of my men are killed, you will be charged a sizable amount for an unnecessary death."

"That's fine." James walked to the grav car that was waiting on him. He entered, and the car pulled off down the tunnel that connected the docks to the rest of the station. James had time to relax for a few minutes as the grav car soared above the city. James noted that the artificial sky was in good working order, and it appeared to be mid-day. However, James noticed that there were no people on the streets. James synced his comm with the station and found that K-19 was set to mimic Earth's time, even the seasons. When James noticed the time, he knew that the station was not working as well as he initially thought. It was four in the morning. James started to look harder and saw beneath the shiny façade, the station was not what it appeared to be.

The grav car lowered on the designated spot for visitors, and James walked up to a warehouse that looked like it was about to fall apart. As James was about to change his mind, someone stepped out of one of the garage bays on the side of the building wearing a small mechanized suit. James' jaw dropped as he saw the suit. It looked to be a newer model of the one Ski wore back on the asteroid, and he wanted one. A distinctly feminine voice boomed from a speaker mounted inside the mech suit. "Mr. Minder, please come around to the side. You won't find what you are looking for at the store front. That's just for the locals. With the amount of money you're about to spend, you get to see our real inventory."

With just one step across the threshold of the roll up doors, James was transported from the dingy, rundown warehouse to a spotless showroom with all manner of weapons, armor, and vehicles. For the first time in a long time, James was glad that he had more money than he could spend in a lifetime. "Today is going to be a good day, buddy." A green light winked on James' comm.

Chapter 14

Bob's Surplus was a front for one of the galaxy's largest and most advanced weapons manufacturers, S&K Arms. The local gangs all used Bob's Surplus as a neutral site to buy and sell low grade weapons. A couple of times in the past, the gangs built up the courage to try and rob the place. The bodies were never found. The gang that the men belonged to was charged a penalty price for all future purchases. This was enough to stop all but the most desperate of gangs from trying anything.

After the first visit, James was allowed to walk in the front door and walk through Bob's Surplus to the *Employees Only* doors, where he had to pass a biometric scan. From there, he walked into the lobby of S&K Arms. The contrast between Bob's and S&K surprised James every time. As soon as he passed through the door, the dingy surroundings and musty smell of used military gear vanished and was replaced by a spotless lobby complete with a light lavender scent in the air. Bex, the very attractive receptionist, greeted him with a coy smile for him and treats for Loki. James had seen this before and was certain that somewhere in the building was an equally attractive Rex, waiting to greet customers that would find him more attractive. But he still enjoyed the attention occasionally.

Today, James was meeting with the mech suit designer to add the final specifications to his personal mech suit. During his first trip to S&K Arms, James asked if Loki could get fitted with a mech Suit as well and was promptly shot down. As smart as Loki was, arming a dog with heavy weapons was something S&K Arms was not willing to do. However, they took a look at Loki's suit and told James that they could create an up armored suit that Loki could wear, and it would be far more protective than his current suit. The price for just Loki's suit was 50,000 credits. James didn't even flinch as he placed his order, he was in the millions.

One of the things that brought down the price of Loki's suit was the code that ran all of the systems in the suit. The S&K Arms tech that was copying the code was so impressed with its quality that he asked who made it so that S&K Arms could buy the rights to the code. James felt much better about the coding that Port had done for him on the fake XS-12's, and as an added bonus, Port was about to come into a large sum of money if he played his cards right. With the same code in both suits, Loki would have to learn very little. While James, on the other hand, would be in a simulator, and once the suit was built, he would spend the next month or two training.

While money was no problem for James, he still lived simply for the most part. Every morning, he woke up on his ship, ate breakfast with Loki, and then walked to S&K Arms, where he spent most of his day in the simulator. He worked with instructors on single man tactics, sometimes even Loki was added to the scenarios. He was a horrible attack dog and would just bark in protest if he was told to bite or jump on someone. However, he was an excellent watch dog and always alerted James if someone was coming from somewhere unexpected. Most of the time, Loki went on walks with Bex around the compound and had become somewhat of a local celebrity.

James knew that sooner or later, Ski was going to come for him. So, he trained as hard as he could. He even found himself enjoying the training, especially when Loki was involved. He also had created several shell companies and was able to purchase a much larger ship that had defensive weapons, shield generators, and some impressive offensive capabilities as well as an experimental A.I.. James created a fake bounty hunting firm and purchased the ship with the company as the buyer. James also had several fake identities created within the shell companies so he could hide in plain sight. With surgery and DNA editing, he would be able to change almost every attribute of his body, James was not worried about his appearance. He was worried about his bio markers. Fooling the system for that was much trickier. However, with enough money, he was able to get hacks into all but the most secure systems.

James watched the news feeds with fervor. He knew Ski would not wait too long before using one of the XS-12's. If he was lucky, Ski would be at his mining facility. If he was unlucky, Ski would live through the double cross, and James' training would be put to the test. The day came when the news feeds exploded with information. A planet was being held for ransom for 5 trillion credits. A previously unknown group claimed that they had a new and powerful weapon that would wipe out an entire planet, for proof, they used James' near miss, an example that they had the weapon. James knew the rest of his life would change depending on the outcome of this event. If the ransom was paid, it would just delay the inevitable. Sooner or later, the XS-12 buried on the asteroid would change the course of events in the known galaxy.

Ski had been messaging James every few days, demanding that James deliver the tablet to him. Soon, the demands turned into threats. One day, while James was walking Loki to S&K Arms, Loki alerted James that they were being followed by flashing a quick flash of red, yellow, and green light in quick succession. James thought that this was part of the training he was paying for. When he did not react, Loki flashed the lights again and bit the sleeve of James' EVA suit to make sure he got the message.

James let the training kick in and made sure to remain on the main street, "Loki, how many do you see?" Two red blinks. "Are they from S&K Arms?" Loki tried to look back over his shoulder and sniff the air. Hampered by his normal EVA, Loki had to do an awkward jump turn to try to get a better look and smell. He finished his circle, and a yellow light flashed on his collar. James pulled up Loki's feed and played back the last 20 seconds. "There you are," James said as two men tried to blend into the surroundings when they saw Loki turn around. "You two are definitely not from S&K, are you?" A yellow light flashed, "Never mind, buddy. We need to go."

James called the Keller group. "Yes, this is James. I seem to have two men following me. I'd like it if you pulled some of your people away from guarding my ship and give me a hand... Yes, I will leave the comm line open." James had to take a longer route toward S&K Arms because he did not want to turn down any of the alleys and walk into an ambush or give his would be kidnappers a chance to trap him.

James was off his normal route by quite a bit, and from the glances Loki got from turning around, James could see that the two men following him were growing more concerned as he strayed further and further off his normal path. As he approached the last major intersection on his way to S&K Arms, the two men made their move. James heard them running the 20 feet to catch up with him. Loki was doing his best to scare the man, and his bark was the fiercest James had ever heard.

James froze as both men drew guns and pointed them at his chest. “We need to have a chat.” the man in the yellow jumpsuit demanded. “You’re gonna come with us, and you gonna shut that dog up, or I will put him down.” James looked at Loki. “I need you to calm down, buddy.” James knew when Loki was pumping his calming hormone into the air and had worked with him so that Loki could produce it on command, and James had just given the command. At first, nothing happened, but then, almost imperceptibly, the gun men seemed to relax just a bit. Their gun arms drooped ever so slightly and their posture relaxed. “Look, if you come with us without a fuss, I promise you that we won’t hurt you or the dog.” The second man said in a conversational tone. He wore the same yellow coveralls, but his were faded almost to white, James noticed he was quite a bit older than his partner.

A hushed voice in James’ ear almost made him jump. ”Mr. Minder, we are two minutes out, keep them talking, and don’t let them put you into any vehicles.”

“Okay, okay, we are good Loki stopped barking, what do say about having that chat right here?”

"We are not the one's you'll be chatting with." It was bright yellow who spoke this time.

"Any chance we could arrange a meeting later? I'm on my way to work and don't want to be late for my shift." James lied. The tension started to creep back into the men's bodies. "What do you take us for, James Minder," the older man sneered. "a couple of fools? We know exactly who you are, and you're coming with us right now." Both men had their weapons point directly at James's chest now, Loki's attempt to calm the situation had run its course.

"One minute, Mr. Minder, you can start to follow them now, we will intercept you before they can take you too far."

"Well, I guess the game is over, lead the way, gentlemen." James and Loki fell in step with the older man as the younger man kept pace behind them. As they walk, the voice in his head calmly told him," On my three count, I want you to drop straight to the ground.... three...two...ONE!" As James dropped, he yelled "DOWN," and Loki almost beat him to the deck. By the time James looked up from the ground, the old man lay on the ground convulsing, and the young man behind him was laying face down with a sizable piece of his skull missing. Blood and gray matter spilled on to the paint and steel of the sidewalk. As the blood spread, the mans yellows coveralls started from bright yellow to deep red.

James had seen death before, but never this close and never so suddenly. The smell is what got him, it was so thick that James could taste it, the copper and musty scents mixed and coated James' throat. There was nothing James could do, and his body revolted, the contents of his stomach splashed across the same steel were red puddle was growing. James looked away just before the two fluids mixed. Still recovering, he felt two rough sets of hands frisking his back and legs, without any warning, he was flipped onto his back, and the process repeated itself.

Once the wound check was completed, he was hauled to his feet and stood looking eye to eye with a set of eyes so blue that they had to be augmented. Surrounding the eyes was one of the most beautiful women James had ever seen. Her blond hair bounced loosely about her shoulders as she helped James up without strain. “Oooo, isn’t that cute? Our little man can’t stomach a little violence.” She cooed. James felt a warm flush run across his face. “That is not a little violence,” James shouted as he tried to break the grip that held him up. “that man is missing half his head.”. The angel of death looked unconcerned, and she let James go, “That’s way less than half, and it’s not even missing,” she walked over and kicked at a piece of scalp and brain matter littering the side walk. “See, some of it’s right here.”. James no longer considered her an angel, she was now a devil in disguise.

Chapter 15

James sat in a chair with a vacant expression, trying to process what happened, a light set of slaps across his cheeks snapped him out of his stupor. "You still with me, I need ya to tell us what you want done with this. Sergeant Keller said as she pointed to the bound and gagged man in the back of the armored grav truck. "What do you mean? How am I supposed to answer that? James asked.

"Answer it with your mouth, or if that is too much, maybe write it down..... Hmmm I got it, you can act it out for us, that way, we can all have some fun. Oooo, we can even take his gag out of his mouth, and he can guess, too." Sergeant Keller stared off into space giggled, "Yah, that would be fun."

"Something is very wrong with you." James started to go on.

"Yah, maybe, well probably.... But that still doesn't change our current problem. What do you want to do with him?" Sergeant Keller looked over shoulder at the man who was drifting in and out of consciousness. "I'll even make it easy for you, we beat the information out of him, we drug him and get the information, or we kill him and be done with it. But, as I remember, you asked us not to kill people on the spot. Aaand if we just kill him, we kinda wasted a pair of flexicuffs, then there was all the effort dragging him into the truck. Anyway, what's it's going to be, Mr. Minder, beat, drug, or kill?"

"You killed the other man without asking, why is this different."

"Because silly, the other guy had a gun at your back, and if we tried to stun him, he could have killed you. Aaand if you die, we don't get paid. Now quit stalling, answer the question."

"Take the gag off and wake him up."

"One beating coming up, and I thought you didn't have it in ya."

"What?" James stammered

"What do you mean what, you chose option one, the beating, right?"

"No, I just want him woken up so I can talk to him."

"Borrrring, and pointless. You're lucky you're paying us so much because you sure are a pain. But fine, here we go." Sergeant Keller walked over to the man and ripped the tape off his mouth, the man flinched but did not wake. She looked at James and arched an eyebrow, asking permission to wake the man. James swallowed hard and gave a curt nod. Pulling the man up by the hair, Sergeant Keller leaned in close and lightly tapped his face with her palm while whispering in his ear, "Rise and shine, sweetie." Nothing happened for a second, before James knew what was going on, the Sergeant pulled back on the man's hair and slapped him so hard Loki yelped.

As if by a miracle the man was wide awake with a delicate bright red hand print on his right cheek. "What happened to not beating him?" James asked

"What, I tried to be nice, but, he wanted to keep pretending that he was asleep, it's his fault, really."

"Fine." James turned his attention to the bound man. "Why were you after me?"

The man sat cheek glowing, not saying a word. "Oooo, I have an Idea," Sergeant Keller said in her most sweet voice "You should say please."

"Look.... guy, we know that you were awake for at least most of the conversation, if not all of it. So, you should know that I'm the only thing stopping you from being murdered in a very slow and unpleasant way, you might want to work with me here. I'll try again, why are you after me." The mans face hardened, and he looked away. Sergeant Keller once again raised her questioning eyebrow toward James, he had barely started to nod when the shot rang out. The scream that followed made James cringe. Loki started barking and ran under the chair James was sitting in. "What in the hell was that?" James shouted over the ringing in his ears, "I thought you were going to hit him or something."

Sergeant Keller was doing a poor job containing her laughter. "I know, you should see your face, I don't know who was more surprised him or you."

James looked over at the screaming man, a large portion of his left foot was mangled and dripping blood on to the floor. Sergeant Keller picked up on his legs, causing him to fall back into the truck bed, as he lay there writhing in pain, Sergeant Keller grabbed the remains of his foot and squeezed with her bare hands until the screaming stopped. "Thank the gods, his whining was giving me a headache."

"Aren't you supposed to ask a question while you're torturing him?" James could smell the blood, and his stomach turned.

Sergeant Keller dropped the mangled foot as she ran her hands through her hair, turning it red, she laughed, "I always forget that." She reached over and grabbed the mans hair pulling him into a sitting position, with her right hand she delivered slap to his left cheek the man screamed as he regained consciousness. Sergeant Keller stepped back, admiring the matching hand prints that covered the man's cheeks. "Hey, Sweetie, I forgot to ask a question before you passed out, could you do me a favor and tell me why you were trying to kidnap my friend?"

The man stammered, "I picked up a contract on the net, we were supposed to hold you until a man named Ski got here." There it was, James had just run out of time. "Anything else you'd like to tell us, I don't know, like when he is getting here, you know anything like that?" Keller asked with a sweetness that never hinted at the monster that she was.

"He never told me, just gave me a comm address to let him know when I had him." He pointed at James.

"Well, thank you, you have been a great help," She turned to James. "Do you have any more questions Mr. Minder?" James was barely paying attention when he shook his head no. The roar of a gunshot snapped James back to the present.

The bound man slumped over in the bed of the truck. The copper musty smell was back. Sergeant Keller was holstering her pistol as she walked out of the room, she stopped and called over her shoulder "I'm hungry, let's get some chow. I'm craving orange chicken you in?"

Chapter 16

James was stunned, had that just happened, had he participated in a cold-blooded murder. He ran out of the garage after Sergeant Keller. James saw her turn a corner and called, "HEY KELLER, WE NEED TO TALK!". Sergeant Keller stepped back around the corner just as James was about to turn the corner at full speed. James bounced off her armor and tumbled to the floor, "I don't know what kinda game you're playing, Mr. Minder, but a little guy like you should be more careful when running after folks." Sergeant said while looking at James, "Here let me help you up." James took the offered hand and was hoisted to his feet with surprising ease. "Hope I didn't hurt you."

"Sergeant Keller, you just murdered a man in cold blood!" James said with disgust. The Sergeant just waited. After a few tense seconds, she turned and started to walk off, completely confused, James reached out to grab her arm and was stopped cold. "If you value your arm, Mr. Minder, I suggest you do not place your hand on me." All the sweetness was drained from her voice, all that was left was the promise of violence and pain. James pulled his hand back as if he had been burnt and stopped in his tracks. "You are smarter than you look," Her tone left ice running through his veins. "Now, is there something you wanted to talk about, sweetie?" The rapid change in personality's was as confusing and the angelic looking woman with blood stained hair standing in front of him. "Well, Mr. Minder, you gonna ask me on a date, or just stand there with that silly look on your face. I am already headed to grab some food you'er still welcome, we can even call that a date if ya like."

"What are you talking about? I wanted to talk to you about how you just shot a man." James explained "I shot him with my gun silly, how else are you supposed to shoot someone?" Sergeant Keller asked while turning around and walking away. "Now are you coming to lunch or what?" James almost reached for her again but decided to walk with her instead. "You know what I mean, why did you just shoot the man? For the gods sake, he was tied up!"

"Look, Mr. Minder," She said with the steel creeping back into her voice "did you research the post he was talking about? Did you know that he was lying to us, did you know my men found the post days ago? Those men were not just going to kidnap you, Mr. Minder. The contract was for them to kill your dog in front of you, then torture you until you gave the location of some 'tablet.' After that, this 'Ski' person had a comm link emended in the post so he could watch you be murdered. Once all of that occurred, the 500,000-credit bounty on your head would be paid."

"How do you know all of this, and why are you just now telling me?"

"Mr. Minder, you rented the most expensive berth for the smallest ship on K-19, you ordered our highest level of protection package, and you are going in the back of 'Bob's Surplus' and spending most of the day in there while the most beautiful woman and man I have ever seen walk your dog around. Did I miss anything, Mr. Minder?"

"Well, no, but that does not answer my question so, so I'll ask again. How and why do you know all this stuff about me?"

"Read the contract you signed with us, Mr. Minder, that will answer all your questions." The hardness disappeared from her voice. "Now come on, sweetie, we got a date." All James could do was follow her like Loki was following him. Once they got to the cafeteria of the Keller Group building, true to her word, Sergeant Keller got the orange chicken, and James and Loki followed suit.

After a few minutes of James reading his contract while poking at his food, he realized that he had paid for all the services that Sergeant Keller already provided. The space-net monitoring the five minutes on call service, the ship guards, all of it, and so much more was included in his protection contract. "If you knew all of this, why didn't you tell me?"

"James, I don't like to talk about work while I'm on my dates. But if I must, how often to you check your messages from us?"

"I get the daily invoice, but I really don't look at it."

"Men! Go ahead and open one of them up and read the complete title for me." James opened the message, and his face went red with embarrassment. "Well, don't be shy, read it to me, sweetie."

"Daily invoice and Sitrep." James mumbled

"Now, is everything I told you in that brief?" James read through the Sitrep and nodded his head.

"But why didn't you just tell me, wait you've been calling me James and not by my last name I don't even know your first name, do I just call you Sarge?"

"No, never, sarge" Sergeant Keller hissed. "My name is Trish, and you can call me that because we are on a date. And this is the last work question I will answer on our date. How many people do you think the buy exact package you got and never look at us, let alone talk to us? The answer is all of them, to include you. When I first started up this company, we tried to give in person daily briefings just for the person we were protecting not to show up or one of their underlings tell us that it was not required. After a while, we found out that the net messages met the intention of the contract so we stopped in person briefs."

"Trish, I'm sorry. I didn't mean to treat you or your people like that. I will do better in the future."

"Don't go getting too sappy on me now, after all, this a just our first date. I know you still want to talk about your current situation, but at least for now, let's just eat and make small talk."

James and Trish talked throughout the meal, and he almost forgot about the days earlier events until he looked at her hair. Trish met his eyes and blushed, she then held her head off to the side of the table and started rubbing all of the dried blood out. After a few seconds of rubbing and scraping, most of the blood was gone, she then produced a brush from some where and smoothed her golden locks back down, and the brush disappeared. "Sorry about that, I forgot." She explained as though it was the most normal in the situation. “Well I guess it’s back to reality, Mr. Minder.” She said while getting up to put her dishes in the scullery.

“Can I still call you Trish?”

“Let’s see how the second date goes first.”

“There is going to be a second date?”

“That, Mr. Minder, depends on you, but I’ll warn you, take me to the cafeteria again, and there won’t be a third.” They started back toward the garage area.

“I’ll keep that in mind.”

“Now for some more good news,” the switch flipped again, “the men we killed earlier today did not send that Ski fella a single comm, but as far as we can tell, at least ten others did. Now for the really good news, not a single one of the others was on this station.”

“That’s great news, Tri, I mean Sergeant Keller.” James exclaimed, interrupting her.

“Let me finish.... Yet.”

“Yet....?” James asked, completely lost. “Not a single one of the other messages was from this station yet. In case you haven’t noticed, Mr. Minder K-19 is not the safest place in the galaxy, and to tell you the truth, if you were weren’t going to BOB’s everyday, you would have been targeted by the gangs in your first week here.” Sergeant Keller explained

“So what do I do now?”

“You finish up your business with Bob’s Surplus, and you move on, as much as I like taking your money, 500,000 credits will bring more trouble than my small team can handle. You need to leave before someone finds you. How much more time do you need at Bob's?"

"My training is almost done, and the suit will be complete in two weeks," James explained

"Good when you are done with your training at Bob's for the day, my guys will pick you up, and you will train with us." Sergeant Keller said with a sweetness that scared James, "You are going to need all the training you can get if this Ski guy is throwing that kinda credits after you."

Chapter 17

James woke early the next day, he kept to the standard 24 hour spacer day, although it really did not matter since the artificial sun was stuck at the noon position in the station. He wondered what the second part of his day would be as he absentmindedly went through his morning routine. James was so preoccupied that he switched his and Loki's breakfast plates and almost took a bite of Loki's breakfast, by the time he realized his error, Loki had devoured James' food and was looking at the dog food in front of James. "Fine," James placed the plate at Loki's feet, "but I am telling Bex that instead of a walk today, you get to go on a run." Loki paused, huffed, and then finished eating.

The news feeds playing in the back ground all started show different versions of the same story, "TERRORIST GROUPS CLAIMS TO HAVE DESTROYED AN ENTIRE ASTEROID BELT TO PROVE THEY HAVE THE CAPABILITY!" as James continued to read, he knew someone had detonated the XS-12 buried next to Ski's mining facility. From what James gathered from the articles, the time had run out for a planet being held for ransom by a terrorist group calling themselves Free Worlds of the Milky Way (FWMW). The terror group detonated the weapon to show the planet being ransomed that the threat was real. The FWMW gave the planet a week to pay the ransom, after that, there would be no more warnings.

An unknown comm with an urgent message interrupted James while watching the feeds, without thinking twice,

James dismissed the message without reading it and went back to watching the news feeds. A pounding on his inner cargo hatch a few minutes later caused Loki to go into a barking fit. James was stunned. "How did someone get past the guards and the outer hatch?' James asked while trying to calm Loki, a yellow light appeared on Loki's collar. James only had the interior door closed because Sergeant Keller had yelled at him when she had dropped him off, and he did not have his security systems set. Maybe her advise had just saved his life, James could have sworn that the plazsteel was buckling with each blow that continued hammering into the door.

James found his comm and the tablet to the XS-12 he kept close as insurance, he activated the deadman switch option and comm'ed Sergeant Keller. A woman roared into his ear, "IF YOU DON'T OPEN THIS DOOR RIGHT NOW, I WILL BE THE ONE TO KILL YOU AND COLLECT YOUR BOUNTY!" Without a second of hesitation, James closed the comm and opened the door. There before him stood the brightest pink set of powered armor that the universe had ever seen, the color only distracted James for a second before the face plate raised and revealed the face of the demon encased within. "HAVE YOU LOST YOUR MIND" Sergeant Keller screamed "WHAT IF I WAS SOMEONE TRYING TO KILL YOU?" James stumbled back and screamed back. "By the way you're acting, I think you might actually be trying to kill me. Would you mind telling me what this is about?" James asked as he disabled the deadman switch before Sergeant Keller did something they both might regret.

“I thought after yesterday, well I thought that maybe you would’ve taken some things we talked about to heart.” Sergeant Keller exclaimed

“You need to explain what is happening because I’m completely lost.”

“What’s happening is you are about to have every bounty hunter this side of Sol searching for you.”

“I STILL DON’T UNDERSTAND WHAT IS GOING ON!”

Sergeant Keller gritted her teeth “Get your EVA suit on, have Loki in his armored EVA Suit, and get in the truck; you have two minutes, MOVE!”

“I won’t.” James was interrupted when Sergeant Keller‘s armored palm smacked his face so hard that that he tasted blood.

“I SAID MOVE!” Sergeant Keller roared, this time, James complied without question. When Loki saw James putting his suit on, he walked over to his armored suit and begin to wiggle his way in. As normal, Loki was done before James, but not by much, Sergeant Keller’s slap had cleared James’ mind of all questions, and his only purpose in life at the moment was to get the suit on within the two minutes. Loki and James both made the time with scant few seconds to spare, “Put your face plates down and turn the face plate to the mirror, I don’t want people seeing your face.” Sergeant called as she turned and headed to the armored truck.

During his last ride in the truck, James had wondered why the cab was so oversized, he did not have to wonder any longer, the mechs suit surround him barely fit within it confines. “Are you going to tell me what is going on now?”

“Mr. Minder, what have you been doing this whole time we ran to the truck and when we’ve been riding in here!?”

“I was wondering what was happening, why?”

“I have half a mind to push you out of the truck and let the bounty hunters have you. This is the last time I am going to tell you read your Sitrep, I am on the radios and don’t have time to explain something you already have the answers to.”

James said nothing, but he did pull the Sitrep up on his HUD. After reading it twice, he understood why Sergeant Keller was acting the way she was, and to tell himself the truth, he was not 100% sure why she was still helping him. The FWMW had placed a 5,000,000 credit bounty on his head, this time, the only stipulation to the contract was that his ship be turned over to them intact and with all contents. James then knew that Ski was alive and the FWMW was nothing more than a money generating enterprise.

“Where are we going?”

“We’re headed to S&K to get your suit, and then we are getting out of here before too many people find out you are here.”

“You know about S&K Arms? I thought that place was a secret.”

“James, look at me. I am in a pink mech suit; do you really think I just found it on the side of the road or something? I’m under the same NDA you are. Now you your suit is complete and they are giving you the last couple of sim chips.”

"So why didn't you just pick up the suit and bring it to my ship?"

"Have you completed the bio imprint and neural link?"

"Well no, we were going to do that next week. I was told that neural link was not a pleasant experience and they wanted to make sure I was able to complete the training before subjecting me to that."

"Not pleasant is an understatement, most people pass out during the procedure and take a week to recover fully. You don't have that kinda time, that's why I am going with you."

"Ok, I guess, I just wish they would have explained it better when I signed up for this."

"Did you even read the contract you signed, it spells out very clearly what is going to happen to you and the possible side effects."

"I may have skipped over a part or two."

"Mr. Minder, if I did not know the lesson you are about to learn, I would throw you out of this truck right now. But I have a distinct feeling that after today you are going to read things a bit more thoroughly."

"Is it really that bad?"

"No, it's so much worse, but the good news is that we are here."

Chapter 18

James was shaking as he was strapped into the medical bed face down, the medical devices whirled to life. "Tell me again why I have to be awake for this." The room was cold and sterile; a typical modern operating room.

"Because, Mr. Minder, to ensure we have the link correctly placed, you have to be awake to provide us feedback. Now hold still, this will be unpleasant." The med tech said the last part apologetically. James' world exploded into pain, and his vision started to tunnel. "James, I need you to stay awake, if you pass out now, we will have to start over." With the promise of more pain, James willed himself to stay conscious. Next came the questions that seemed to never end, and the only thing that kept James focused was the shooting pain that coursed through his body randomly.

After what seemed like an eternity to James, the procedure was almost complete. "Okay, Mr. Minder, this is the last part, the pain is over, but this next part will probably be one of the oddest feelings you have ever experienced, but you will get used to it. Are you ready?" James' body was covered in a sheen of sweat, and he was still recovering, "Ya, go ahead as long as it doesn't hurt, I think I can handle it." James gasped as he felt two distinct metal prongs sliding into the base of his skull on either side of his spinal cord as the neural bridge was temporally placed to test the neural link. If he was not strapped down he would have jumped when they clicked into place. All of a sudden, James' vision was augmented, he had so many overlays he could barely see through them. James started to get a tension headache from the overload of information. Then the voice started, "James, James, James, Loki, James" The voice was monotone and sounded like an unprogrammed universal translator.

"James," the med tech shook him.

"James, I need you to tell me what is going on."

"I am hearing a voice in my head calling my name over and over again, I can barely see, too many icons in my.... eyes I can't concentrate.

"It's okay the visual is perfectly normal, if you want to clear your vision, concentrate on the lower left corner there is an option menu. Open by thinking about it."

"Okay, another overlay popped up." James stammered

"That's okay, find the option that says clear HUB and think select." James' vision cleared and the only strange thing was the voice in his head saying his and Loki's name over and over.

"How is that, feeling better? The med tech asked

"Ya, but I am still hearing the voice."

"Yes, I figured that out. I'm going to unstrap you now and show you where the voice is coming from."

James slowly rolled over and then sat up. "Okay, doc, where is the voice coming from?"

"You see the window," James nodded. "Do you see who's in the window?"

"No, I just see Loki."

"Think open comm to Loki, then say hi."

James did as he was told, and the voice in his head changed to machine gun fire of "Loki hear James." over and over. Loki was jumping at the window like James had never seen him move before. "Loki has a neural link?" James asked, he then thought, "Loki, I need you to calm down, you're going to hurt yourself." Loki calmed down, and after a few minutes of James telling Loki everything was fine, James asked Loki to not comm him for the next few mins, the doctor was talking to him. He watched Loki's head droop and tail stop wagging. "Buddy, I'm sorry I promise I'll talk to you soon this is just a lot for me to take in right now." Loki's comm channel blanked out on James' HUD, and now his vision was completely clear, and no one was in his head.

"Mr. Minder, normally we would wait a couple of days to do the bio imprint on your mech suit, you know, to allow you to recover, but Sergeant Keller insisted that we bio imprint you today and that you will no longer be training with us. We loaded your suit with rest of the sims you will need to complete your training. Are you ready to move next door and begin the imprint?" James edged off the bed he had just been strapped to and tested his legs, not fully trusting himself not to just collapse. "Ya, let's go." James walked toward the door on shaky legs.

By the time he got next door, he was feeling much better, then he walked in the room and saw Sergeant Keller standing there in her pink armor, and he got really worried. "What.... What are you doing here?"

"Two reasons, first I need to hurry you up, we caught some chatter over comm's that you were spotted. Right now, there is a group mercs headed this way to collect on your bounty."

"Okay, what's the second?" James asked as Sergeant Keller pushed him into the back of his open suit.

"Ya know how bad the neural link hurt going in, well, that's how good it's going to feel your first time plugging into your new suit." Sergeant Keller whispered in his ear and pulled out his neural bridge. James's knees would have buckled, had Sergeant Keller not activated the close function of James' new mech suit. He felt the prongs from his suit enter into the back of his neck as the suit begin boot up for the first time, a sense of euphoria washed over James, like he had never felt before, as more and more systems came on line his senses heightened and James could feel the hard points holding him up. He could see in a wider spectrum, and he would swear that he could smell the lubricating grease that was applied to his suit's joints. "This is so much better than the simulations." He exclaimed to no one.

"James, you need to come with me." Sergeant Keller's voice sounded in his head and off to his right side at the same time. "We have to get you armed before we leave, and we are leaving yesterday." James opened the comm to Loki by thinking about it, "Loki, I need you to get in here and follow us, do you understand?" A very familiar voice came over the comm the same time, a green light lit up. "Loki, yes, understand." James linked the comms just like in the simulators, now Loki and Sergeant Keller could communicate with each other. "Lead the way, Sergeant," James spoke as soon as Loki entered the room.

Sergeant Keller had just run outside the room and into the hall that would lead them to the armory when the first rocket smashed into the complex. James was not so lucky he was still in the room when the rocket exploded almost directly over him. He was knocked off his feet and slammed into a wall by the explosive force. The explosion would have killed James if it had happened two minutes earlier, but as far as James could tell, he did not even scratch the paint, however, the wall he slammed into would forever have a James shaped imprint in it. Back on his feet and running faster than he had ever run, James caught up to Loki and Sergeant Keller in the armory. She was already loading a wicked looking rifle, as she finished, she threw the rifle to James. He saw the vector's, and spin rate of the rifle on his HUD as it came at him, distracted by all the new information he missed the rifle, and it bounced off his head and chest. Luckily for him, he was able to catch it before the rifle hit the ground, Sergeant Keller gave him a withering glare and loaded her own rifle.

The building continued to shake as fire from the mercenaries poured in. After James had a full combat load out of ammunition and explosives, Sergeant Keller comm'ed James and Loki, "Loki, Stay behind us and let us know if you see anyone coming from behind us." A hauntingly familiar voice replied, "Loki understand."

"Loki, I need you to change your voice output I cannot have two James's in my head." Sergeant Keller exclaimed, despite their predicament, James laughed into the open comm. "That's why it sounded so familiar." "Loki, okay?" Loki's voice was completely monotone again, but James could feel the sadness.

The trio made their way to the garage bay. On the way, James saw Bex and the guard that approached them the first time they came to S&K Arms, as well as several others in full combat gear, and headed in the same direction. A comm request popped up on James' HUD. "Mr. Minder, we will provide a few seconds of cover fire before we lock down the building." Sergeant Keller cut in, "Concentrate your fires to the north we will break south. We should only need a few seconds to break through the ambush." James recognized the stack formation at the door from his simulator training and got into position behind the sergeant. "Loki here." Loki's monotone voice said.

The garage opened and they all spilled out into the courtyard. James froze for a split second when he saw the amount of people shooting at him, he was quickly snapped out of his stupor when a string of rounds slammed into his chest armor. Now he was sure the paint was scratched, James started to run to cover as he lifted his rifle. A crosshair appeared on his HUD over an unarmored mans head. James pulled the trigger.

Chapter 19

James' world slowed to a crawl, he saw the charged plasma core round leave the muzzle, he saw the compressed gas expanding in the bullets wake, and he watched round as it entered the bridge of the man's nose. The rest of the world disappeared as James saw the light leave the man's steel grey eyes, there was no body flying through the air like in the vids, no fountain of blood, just a small hole. The man's head barely snapped back as his body dropped behind the grav car he was using as cover. James realized that he had not even broken stride as he took a man's life. He knew that those eyes would haunt him for the rest of his days.

Real time threatened to crush him as it went seemingly from a crawl to a dead sprint in an instant. James almost became overwhelmed with the shear amount of information and events happening around him. There was a light sting on his left butt cheek and instantly the calming effect of a sedative along with the clarity and focus of a massive adrenaline boost. "I gave you a combat cocktail help you through the next few hours, now move faster, or I will shoot you myself and collect the bounty." Sergeant Keller yelled in his head.

James quickly fell back on his limited training and begin to sprint faster than he ever could on his own. He was keeping pace with Sergeant Keller when he realized what break through the ambush meant. They were running straight at the mercenaries firing at them. Sergeant Keller did not even seem human to James, as she ran, the only time she stopped firing was to throw explosives. As they cleared the blockade that the ambushers were now cowering behind, James saw the trail of death and destruction they left in their wake.

"Loki get out ahead of us about 100 yards and let me know if you see anyone that looks like the people we just fought. Do you understand?" Sergeant Keller asked with out loosing a step. "Loki, yes." The monotone voice said while an armored fur missile shot past both James and the Sergeant then slowed to their pace at about 100 yards in front of them. "I don't like you using Loki as bait." James protested.

"Hey, sweetie, look behind you and tell me where most of the danger is coming from." James did as he was told and noticed that they where still taking fire as the merc's regrouped and began to chase them. He nearly tripped as he was looking backward, "What are you doing?" Sergeant Keller's voice rang in his head.

"I am looking back like you told me to." James regained his footing and closed the gap with Keller.

"You look with your rear facing camera not your head, stop acting like it's your first day."

"IT IS MY FIRST DAY!!!" James roared.

"Oh ya..... Well then, let that be lesson one, use your cameras, not your head."

"Well, what's lesson two?" String of rounds flashed by his head as Sergeant Keller returned fire on a group merc's closing in on them.

"I don't know, maybe when in a fire fight, you should always shoot back."

"Good lesson." James pointed his rifle backwards over his shoulder and fired a couple of bursts.

"What are you shooting at?"

"I don't know, maybe the guys chasing us."

"You might want to use your CAMERAS, sweetie, that is unless that building you just shot is a merc in disguise."

"Remem-" James was cut off as a trio of rounds hammered into his back.

"Shut up, run, and shoot. Make sure you are doing it in that order, too." Sergeant Keller explained at a full sprint.

The running gunfight lasted all the way to the ship berths until, "Loki 10 front." Loki's emotionless voice called the groups attention forward of them. There, they saw three of the men from the Keller group strewn across the entrance to the ships berthing. "That explains why they did not answer my comm's." Sergeant Keller said without an ounce of emotion. Still, at a full sprint, she told Loki to get in between her and James. Now, the fire was heavy enough that the trio had to take cover in a near by building.

Sergeant Keller kicked in the door to what appeared to be a maintenance shop of some kind. The three story building would have to be where they would make their stand. "James, Loki, second floor now." She made a quick sweep to make sure that no one was there and to find all the access points to the second floor. With only one access point she told James to shoot anything that tried to come up the stairs, she continued to the third. "Where is the rest of your team?" James quarried.

"Well, there are those three on the ground you see there, and then I am pretty sure that the rest of them are currently shooting at us." James heard a string fire from up stairs "Yup, they are definitely shooting at us. Well they *were* shooting at us," She stated as matter a fact. "Well what do we do now?" James could feel the panic begin to rise. "Well there is only about ten of them left by my count, so here is what we are going to do."

Sergeant Keller explained her plan, and James shocked by its simplicity, "There is no way that this is going to work, it's going to get us killed."

"Well," she asked sweetly, "what's your plan? That's right, you don't have one, besides if this doesn't work, all of this will be someone else's problem, sweetie."

"Fine, but if I die and you live, I will haunt you."

"That's ok, you will have plenty of company if you do." The Sergeants voice trailed off. "Anyway, it's going to work, so just start screaming on all open channels next." A pair of rounds bounced off James' Helmet giving him an instant headache. "Now, start screaming now and remember the plan." James open all comm channels and started to crying and begging for his life, he explained that he was injured and Sergeant Keller was dead. After several minutes of this someone finally comm'ed him."Throw out all of your weapons and come to the front door." The gruff voice demanded. Havoc pleaded, "I can't see, the last shot knocked out my helmet cams, I don't know how to get it off, come help me. I pay whatever you want, I'll even give you the mech suit."

That got the merc's moving. Almost all of them started running at the front door and into James field of fire. Instead of firing he pretended he was blind and was trying to get his helmet off, that was all it took for the rest of the merc's to charge in to the door to be the first to get to James. As the last merc came through the door and into the staircase "NOW!!" Sergeant Keller's voice yelled over the comm's, a startled James jumped into action. He stopped rolling around and acting like he was hurt and triggered the detonator icon set in his HUD. The staircase blew out killing three of the Merc's instantly. In the confusion, the bright pink figure that dropped from the third story went unnoticed until the mono-blade entered the first man's back. With the single molecule thickness at its edge, the mono-blade could slice through most armor with ease. James did not even fire a shot as the pink demon tore through the men with the speed and ferocity that James would remember for a very long time.

When she was done, nothing except her was alive on the first floor. "Loki one behind." Loki stated over the comm. James turned to see a man climbing through the window. Loki jumped at the man, trying to knock him out the window, instead, the man drew his own black tanto tipped mono-blade and slashed a cross Loki's chest and front legs. James screamed in horror as the blood poured from the stump where one of Loki's front legs used to be. Without giving it another thought, James raised his carbine and fired burst into the man's chest. Over the comm came a weak, monotone voice. "Loki hurt."

Chapter 20

James ran to Loki's side and froze. He had taken a basic first-aid course, but that was for humans. He was completely out of his depth. "Keller, I need help! Loki's hurt bad, get here now!" Sergeant Keller stopped her search for enemies and climbed up what was left of the stairs. When she saw Loki, she immediately shoved James aside and pulled out her field med kit. "James, I need you to give me control of Loki's suit now." She said, trying to remain calm. James fumbled through his HUD menus until he finally saw the menu option to share control of Loki's suit. He shared control of the suit as fast as he could.

Sergeant Keller was spraying a bio foam into Loki's chest and onto the leg that was still attached. As soon as she had the bleeding stopped, she whispered, "I'm sorry, Loki." James begin to ask a question when he was shocked to his core. Loki's suit began to constrict around the severed limb. Without warning, the suits internal seal on the shoulder slammed shut, severing the remaining stump of flesh and bone of his right foreleg.

"What are you doing? Why did you cut off his leg?" James cried.

"Mr. Minder," Sergeants Keller's voice cut like a mono-blade, "if you don't get it together, Loki is going to die, the next time you will see him will be on the rainbow bridge." The statement hung in the air. James swallowed his anger and frustration, "What do you need me to do?"

"I gave Loki the strongest sedative he had stored in his suit, it should keep him out until we can get to your ship, but we need to hurry. Grab that man's mono-blade and sheath he won't be needing it any more " James just nodded his head, "What else do I need to do?" James repeated as he completed gathering the blade and sheath.

"Well, since you are such a great shot," she smirked, "You're going to carry Loki. I will be on point, and we will sprint to your ship. I think that we got all of 'em, but I would be surprised if we get out of here without more shooting."

"What do I do about his leg?"

"Leave it. We'll get him a prosthetic when we get to the ship. Are ya ready?"

"Yah, but I will need help getting him down what's left of the stairs."

James and Sergeant Keller carefully maneuvered Loki's unconscious form down the broken stairs. Once they made it to the remnants of the front door, Sergeant Keller signaled James to stop. The pink mech suit made its way into the middle of the street. As she stood there shouting for the next person to come meet, James was in awe and terrified at the same time. Sergeant Keller stood there, blood smeared across her armor, mono-blade short sword in one hand, rifle in the other, challenging the entire space station to try their luck.

After a full minute, Sergeant Keller gave up. "Well, come on, sweetie, I guess no one has the stones to come out and fight a girl." She carefully sheathed her mono-blade and waved James over. When he got there with Loki's unconscious form cradled in his arms, he asked. "I thought we were going to run for it. What happened?"

"Those two snipers happened." She tilted her head back, indicating that the snipers were in the building directly behind her. She casually rested her rifle on top of her shoulder with the muzzle facing behind her, with her trigger on the finger the whole time. "What are you doing? We're going to get shot." With her free hand Sergeant Keller started over exaggerating every movement like she was mad about something, but her voice had returned to her sweet and calm cadence. "I'm using you as bait, silly." Before James could protest, he saw the two snipers pop up in the windows they were watching from. James started to yell a warning but before he could get the first words out, two bursts roared from Sergeant Keller's rifle. "That's how you use your cameras." The two men dropped back into the rooms they were hiding in, one missing a large portion of his face, and the other's chest looked like as though it had been caved in. "Now we run," She did not wait for James, "the second man's body armor stopped my rounds, he may try again." James was using all his augmented speed to catch up with Sergeant Keller. He noticed his suit was a newer model than hers as he closed the gap and, for the first time, noticed how worn her armor really was. "STOP RUNNING IN A STRAIGHT LINE!" She roared, and James knew why he closed the gap so fast she was changing directions every couple of steps, and he was just running straight ahead.

"James, open your ships berth and start your ship if you haven't already." He had not. As they approached the berth doors, shots began to smack and burn into the wall in front of them. "Keep moving! You can only take a shot or two before that r plasma core round will make it through your suit!" James at least had the sense to close the doors behind him.

"Get in the ship now." Keller turned just inside the cargo hatch and pointed her rifle back toward the door. "Move, James. It's time to leave." She fired at movement off to her left, and another mercenary fell to the ground. James cleared the door and set Loki on his bed, then ran the few steps to cockpit. The hatch closed, and Keller dropped into the seat next to him. "Move this ship before anyone else gets a bright idea."

"I can't undock the ship. Control has us locked down."

"Use the emergency undock and then arm the point defense cannons. I'll talk to control."

"What are you talking about? There is no emergency undock, and I don't have any ammo for the point defense cannons. I was going to pick that up before I left here."

"I swear to the gods, James. Please tell me you read the manual for this ship."

"I glanced through it. The AI does most of the things I need, so I didn't see the point."

"JAMES, THIS IS THE POINT, WE ARE LIVING THE POINT!" She took a breath to calm herself. "Talk to control and tell them to unlock us, or we are going to shoot our way out."

"What are you going to be doing?"

She spoke through a clinched jaw. "I am going to read the manuals so we can get out of here." Seconds later, the point defense turrets came online, and the ship was released from its moorings.

"Good job, James. I didn't even have time to break the lock. What did you do?"

"Well you know how I have quite a bit of money, right. Well, now I have a little bit less."

The little recon ship shot of the station at a reckless speed. As soon as they were able, they activated the que-drive. Once in safely in their warp bubble, Keller stood up stretched. "Well, we have lots of work to do and the combat cocktail I gave you is still going to keep you up for a couple of hours. So, we better get started. By the way, which one is my room?" James was caught completely off guard. He had a passenger on his ship, he had never had a passenger before. "This is going to be interesting." He muttered under his breath.

Chapter 21

"It's the one on the right, and how long are planning on staying?" James asked.

"Wow, you haven't even bedded me, and you're trying to get rid of me," Keller said playfully.

James' face reddened, "That's not what I meant. I just never had a house guest, or a ship guest, and I'm not sure how this works."

"I know what you meant, but think about this. I just lost my entire livelihood, the men I trained tried to kill me, and the only worldly possession I have is the mech suit I am wearing. So, for the foreseeable future, you are stuck with me. If there are no other questions, we need to get started. When the combat cocktail wears off, you are going to be useless for at least 12 hours."

"Nope, no questions."

"Alright, I am going to peel off my Mech suit and hit the sonic cleanser. When I get out, you had better have fabricated me a jump suit, and it had better be pink." Keller walked into her room without a glance back.

"Fabricator? I don't have a fabricator on the ship. I had better just give her one of my jumpsuits. I don't want her mad at me." James muttered to himself.

James hung the jump suit on her door and headed into his room, his heart broke when he saw Loki laying on his bed. James panicked until he saw Loki's chest rise and fall.

He had stopped bleeding, and James could see through Loki's Helmet that he was still unconscious. "It's okay, buddy, we'll get you all fixed up." James stripped out of his mech suit and stood it in the corner of his small quarters. As soon as his neural link was disconnected, James felt like the world muted. He lost all of the HUD information, and he was no longer physically connected to the comm network. James felt more isolated than he ever had.

The sonic cleanser blasted his body with sound waves and water to wash away the stress sweat. James stood in the cleanser longer that he ever had before, trying to wash away the events of the day. The door to the cleanser was abruptly ripped open without warning, and there before him stood the naked form of Trish Keller. "What part of 'pink jump suit' did you not understand? I don't want your hand me downs. Now get out and show me how to activate your fabricator." James was so shocked that someone had interrupted his shower it took him a second to realized they where both naked. He quickly tried to cover up and stare at her at the same time. A snap of her fingers brought his wandering eyes back to hers. "I hope you got a good look because it's the last time, sweetie. Now get out of there and make me a jump suit."

James fought to keep his eyes from wondering too far, "I....I don't have a fabricator on the ship." he managed. Keller placed her fists on her hips with enough force to make her chest bounce. As James stared, he began to notice all the scars crisscrossing her body.

"James, if I have to tell you one more time to keep your eyes up, you will regret it. I saw the fabricator when I was running in here from the ships berth. It was right next to the med pod. Now, get out and make me some clothes."

"Sergeant Keller."

Keller interrupted, "Don't call me that. Sergeants have troops to lead. I don't have that any more. Call me Trish or Keller."

"Um, ok. Trish, can you go put on the jump suit I left for you for now. I need to figure out the fabricator... Wait, you said I have a med pod? Can we put Loki in it? Will it fix him?" James sounded hopeful.

"I don't know. I was going to look into that after I got dressed, but let me guess, you didn't read the manual so you have no idea, right?" James looked sheepishly down at his feet. "No, I didn't, and no, I don't." Trish turned and called over her shoulder, "We have some serious work to do. You figure out how to operate the fabricator. I will work on getting the operating suite running. But I need to be added as crew on the ship's computer so I can access the systems."

James watched as she left. He was stunned, not just by her stunning figure, but the amount of damage her body had taken. James could not see a part of her body that did not have some type a scar or discoloration. "Now, James." She called. By the time he got dressed and added her to the crew on his room terminal, Trish had the med suite pulled from the roof of the cargo area and moved the stacked XS-12 along the back wall. James panicked, "Trish, did you move my cargo?" He asked, hoping that she did not know what she had just moved.

"Ya, had to turn off the gravity just to pick them up. I wanted to throw them into space to give us more room, but thought better to talk to you first. Now pull the fabricator down and get started on the list I'm sending you now." James' comm beeped and showed a long list of items Trish wanted made. James opened the other terminal in the cargo area and found a manual on how to operate the fabricator. Operating the machine was so simple, James became ashamed that he had no idea that he even owned one, let alone use it. This one machine made would change the way he lived. He no longer had to buy things, just buy the files to make them himself.

Trish's list was inputted and would barely put a dent in the raw stock that the fabricator used to produce all manner of items. James asked Trish what all of the parts were for. She explained that it was things that the operating pod needed to preform Loki's surgery. The fabricator whirled to life and layer by layer parts began to appear. James discovered that he had one of the top of the line fabricators made. It could build items at the molecular level, or at its top speed, it could fully fabricate ship parts in under an hour. Trish's list would take about two hours to complete, most of the time would be spent making Loki's prosthetic leg. Apparently, the files Dr. Robinson sent had a whole assortment of items for Loki including several medical devices and medicine.

James and Trish carefully placed Loki in the med pod and began to remove his suit. Most of the bio foam held, and the only blood was from the places in the suit where the blood had not congealed yet. Once Loki was prepped and the pod was closed, all they could do was wait for the parts to print and load them in to the med pod when instructed. James had about six hours before Trish predicted he would crash. He did something he never thought he would do while we waited, he started to read the ships manual.

"Hey James, I have a question?"

"Go ahead, Ser... I mean Trish." She gave him a mean side eye.

"We are kinda stuck together for a bit, right?"

"I think so."

"Okay, so when are you going to tell me about the XS-12's you have stacked in your cargo bay?" She paused, letting the question hang in the air. After several uncomfortable minutes she continued, "I can tell by the stupid look on your face you are trying to come up with something. I will tell you here and now that I have little time for liars. So, how about ya go ahead and tell me the truth." The look she gave James left no room for misunderstandings. James launched into a long narrative about his adventure up until the point met Trish. He did skip over the part about why he bought XS-12's, as well as the part where he almost blew up a planet on accident. "And then you know the rest."

"I believe everything that you said. But sweetie, you left out the part about why you bought the munitions in the first place. Oh, and the part about the explosion in the Jin system. How about you be a doll and clean up that part of the story for me."

"Well, Trish, would you believe that I accidently detonated that Core Buster, and almost died getting it off the planet." Trish's piercing stare bore into James until the uncomfortableness made him squirm. Just as he was about to speak Trish burst out laughing to the point she had tears in her eyes. "Oh dear gods, James, if it was anybody but you, I would have shot you for being a liar. But you... Oh, I bet you didn't read the manual for the Core Busters either." Trish was now gasping for breath between fits of laughter. James' face reddened in anger and embarrassment. "I'll have you know I almost memorized the manual, and I went so far as to memorize the 24-digit code to arm the one that blew up."

Trish stopped laughing, "Well then, James, how did you *accidentally* blow up the deadest weapon ever invented by mankind?" Tension permeated the air between them, James swallowed hard. "I was wiping sweat off the screen and accidentally activated the Core Buster." This time, Trish was on the floor clutching her stomach while tears rolled across her face. Her laughter was so contagious that even James started to laugh despite her mirth being at his expense.

After the laughing spell was over, Trish sighed. "I have decided that I don't want to know why you got the XS-12's," she started to chuckle but managed to stop herself from going into a full blown fit again, "My tummy's already going to be sore for a week, I cannot laugh anymore." She took a large breath, regaining her self-control. "Well, now that is out of the way, what are you planning to do with them now?" Now James smiled, "Well Trish, I'm going to kill myself. Care to join me?"

"This had better be part of some elaborate plan, or I can save you the trouble." Trish patted her holster containing her pistol. "I have a plan." James said quickly, remembering the last time Trish drew her pistol, "But it requires everyone thinking that I'm dead. And since you are hanging out for a while, they need to think you're dead, too."

"Alright, James, let's hear this plan of yours."

"It's simple. I bought the largest ship that I could crew by myself. I made sure it has offensive weapons, and I'm going to blow up this ship, leaving enough evidence to make everyone believe that I'm dead."

"James. Sweetie. That is the *dumbest* plan I have ever heard, and I was in special ops for 50 years, so I know dumb plans."

"I can explain it better. I have created several deep fake personas. Those whose lives I am going to step into. One of them is a bounty hunter, who, by the way, is the one who bought the new ship. I am going to blow this ship up and collect the bounty on my head." James smiled smugly.

"Is that it?"

"No, I also moved all of my money into several shell companies that I can access when *this* me is killed."

"That's better, James, but what about your suit?"

"What do you mean? My suit is fine."

"You're right. Your suit. You do realize that the suit is encoded to you, and wherever you go, it will be registered as being you. How are you going to explain that?" James no longer looked smug.

"I guess I will just get it re-registered," James hoped.

"Nope, anywhere you take the suit to get reworked this *Ski* will likely have informants. We're going to have to blow up the suit. Better yet, I am going to blow it up and kill you at the same time in my new ship."

"I'm not sure I like where this is going, Trish."

"Consider this the price of my cooperation and silence. This is the new plan." Trish laid out her plan in much more detail than James had ever put into any of his planning. When she was done, James was shocked at how much he had not thought of. He was also happy Trish was on his side, even though she scared him.

"Alright James, I think that about does it. Do we have a deal?"

"Yes, we have a deal. But I think I'm getting the raw end of this. And besides, I really like my suit."

"It's okay, James. You're rich, you can just buy another one. And I know just the place."

"Fine, but I don't have to get a different neural link, right? I'm not sure I could do that again."

"No, no new neural link. That was a one-time thing. Now, you need to get to bed by my count, you have about 30 minutes before you crash hard for the next 18 hours."

"What are you going to do? And wait, why aren't you going to crash?"

"Sweetie, I didn't take a cocktail. For me, that was a normal Tuesday. Now go, I will keep a watch on Loki."

James did not feel a bit tired, but Trish had not lied to him yet so off to bed he went. He went to his room and stripped down to his shirt and shorts, then laid in his rack, waiting for the crash to come. From the cargo bay, he heard Trish call, "Hey James, I look forward to killing you soon. Sleep tight, sweetie." Before James could come up with a witty retort, his body started to ache all over, and his mind turned to mush. Fighting the exhaustion was never an option, James fell into the abyss.

Chapter 22

James' sleep was haunted by a set of lifeless steel grey eyes staring into James' soul. At the time, James could not remember seeing all of the details of when he killed his first man, but now, in his dreams, he remembered the color of the grav car, the smell of the ionized plasma, and the sound the round made when it entered the mercenary's skull. James was stuck in a loop every time the mans eyes drained of life, the dream would start over.

When James did finally wake, he had just enough time to make it to the waste recycler and expel every bit bile in his stomach. His body somehow ached like the day he rescued Loki and gotten beat for it. Somehow, James managed to wash his face, pop cleanser tab for his teeth, and then make his way out of his room. As he turned toward the common area, James saw Trish now in bright pink coveralls working at the med pod. James shuffled toward her, "How is Loki?"

"By the gods, you stink, go get in the cleanser you are making my nose burn." James pulled his shirt over his nose and inhaled deeply, now his nose burned and his stomach threatened to revolt again. As quickly as he could, he made his way to the sonic cleanser and through shorts and shirt into the sanitizer. James let the water and sound waves wash the nights sweat away. "James, Loki is fine." James jumped at the surprise intrusion. Trish continued,"Did you know that he would eventually grow the leg back? Any way I turned that gene off, it was going to take a month or longer to grow back on its own. And sweetie, we both know we don't have that kinda time."

"So, if you turned the leg growing gene off, am I just going to have a three legged dog?"

"No, silly I made him a robotic prosthetic that will fit in his suit."

"Isn't his suit ruined I mean, it did almost get cut in half?"

"James, you really need to read the contracts you sign, you bought the rights to all the repair schematics for both or your suits. Its already being repaired by the fabricator as we speak." James had forgotten about his nudity and, shut off the cleanser and flung the door open. "So that mean's I can keep my suit we can just reprogram it in the fabricator." Trish just stared James right in his face until he realized that he was completely naked. He did not want Trish to think she had gotten the better of him again, so he calmly pulled his jumpsuit off the bed where Trish must have laid it. As he put it on, he asked. "Well, Trish, that means I get to keep my suit, right?"

"No, James, when you order a suit like that, they only make a very limited number of them, so because of that, you will need something a little more off the shelf. Also, the fabricator cannot rewrite the code. That is locked behind so many lines of encrypted code that we would never figure it out." James hung his head.

"The good news is that Loki will be out of the med pod in the next hour or so."

James' spirit lifted, he put on his boots and squeezed past Trish on his way to see Loki. As he walked to the cargo bay, he noticed all of the XS-12's stacked against the exterior door, the rest of the bay was filled with two machines. "How in the world did I miss this, are you sure these have been here the whole time."

"Once again, read your manuals." James stood by the control panel of the med pod and looked inside the clear plazsteel dome covering the operating bed. Loki was laying on his left side, from his right shoulder, a black and chrome leg was supported by a robotic arm. Several other arms manipulated the flesh around the attachment point on Loki's torso. James looked at the control panel and saw that the approximant time remaining was 43 minutes, then he looked at the time since the procedure began, 25hours and 13 minutes. "I've been a sleep for a whole day? Why didn't you wake me? Why am I still tired?"

"Sweetie, you slept a whole day because of the combat cocktail, your body pays a high price for that kinda clarity under that much stress. I didn't wake you because you think you feel bad now, just imagine if I did not let your body complete its rest cycle. And last but not least, you are still tired, and I imagine sore because just over a day ago, you put your body through seven different kinds of hell, if there no more stupid questions, let's get some grub, and let's make Loki some too, he is going to be ravenous when he wakes up."

They ate in silence until James pulled out his comm and expanded it to its maximum size, he cleared his plate, place the comm on the table, and began to read the ships manual. Trish snorted out a laugh. "What are you doing?"

“I doing what you told me to do, I‘m reading the manual.”

“What ships manual are you reading, and why aren’t you using your HUD to read it?”

“I’m reading this ships manual, and” Trish cut him off.

“The ship we are about to blow up, you are reading the manual for the ship that is going to be scrap in less that a week. Smart.” James felt the embarrassment welling up in his chest and his face reddening.

“That is a good point.” James switched the manual from his current ship to Trish’s new ship even though it graded him to think of it as her ship.

“Why aren’t you using your HUD?”

“What HUD? I haven’t had a HUD since I took off my mech suit!”

“Where is your neural bridge!?”

“I don’t’ even know what my neural bridge is!”

Trish took a deep breath “Right before you put on your mech suit, you had to remove your bridge, what did you do with it, James.”

“Trish, you took out the thing on the back of my neck, is that what you are talking about?”

Now Trish turned red. “I’ll be right back.” Trish went to her room and returned a few minutes later. With out saying a word, she walked up behind him and slapped the neural bridge in to the two holes in the back of his neck. James winced from the slap and the odd sensation of the prongs clicking into place. A HUD popped up in his vision and linked automatically to his comm.

"So, way aren't you using your HUD, sweetie?" Trish asked sweetly

"Funny, but to be honest with you, this is nothing like the simulator. I'm having a hard time navigating the menus. I think using my comm is the better option for now." Trish acted before James even thought about reaching for his comm, in one lighting fast sweep of her hand, James' comm was gone. You will get this back when I feel that you can use your HUD with the skill required of a mech suit operator."

James stood up, body still aching. "What happens if I just take it back." He said in a playful tone.

"Oh sweetie, I wish you would" A ferial grin spread across her face, and James' blood ran cold.

"On second thought, I will practice with the HUD." Her smile disappeared into disappointment.

"Tease."

After playing with the menus for a few minutes James found that he was able to navigate much easier when no one was shooting at him. After half an hour, he could move through the menus and sub menus while walking around preforming other tasks in the ship. While James was trying to do more complicated tasks while scrolling through his HUD, a comm request flashed in his vision. James focused on the icon, and an audio channel opened, "Loki scared, Loki hurt, Loki hungry." the monotone voice hurt James' soul. "Loki, I'll be right there, and we will take care of you."

James ran the few steps it took to get to the med pod, Trish was there opening the pod and talking to Loki. "Okay, Loki, you are all better, but you are going to be sore for the next day or so."

"Loki, understand."

"Loki, buddy, I want you to pick out a voice. But it can't be mine or Trish's." Loki stared at James with his big golden eyes for a few seconds before a new voice popped into his head. This time, the voice was deep and gravelly, James recognized it as one of the men that stood guard outside of his ship. "Loki understand." James looked to see if Trish recognized the voice, if she did, she did not show it. Trish spoken in James' mind without opening her mouth, "I like it, Loki, you sound super tough." Loki's tail began to wag.

James and Trish helped him down from the operating bed once all four paws where on the floor, Loki dropped his head and sniffed his new black and silver leg, he then picked it up and set it back on the floor several times, testing its response. Once he was satisfied, he begin to prance up and down the main corridor of the ship, gaining confidence in his new leg. Loki spoke into the pairs mind, "Loki up." James was so happy to hear feeling back in his voice that he missed what he said. "What was that, buddy?"

Trish killed the gravity to the ship, and the three of them begin to float. Loki pushed off the floor and was bounding from walls to ceiling and back to the floor. "Loki, can you walk in the ceiling for me?" James questioned. Loki bounced off a wall and stuck to the ceiling, James noticed his right fore paw was not sticking to the ceiling like the other three. "Loki no hurt, Loki no scared, Loki hungry." James laughed.

“Well, Lok, I can’t feed you up there why don’t you come down here, and I will turn the gravity back on so you can eat.”

With a week left in transient, Trish gave the specs to the suit he was going to get and drilled him on the new ships operating manual. Loki was back to normal, but every night, he switched who he slept with, at first, Trish tried to kick him out, but that only lasted until the gravelly voice in her head started to cry. Trish had never heard anything so sad and was forced to share her small bed with Loki just to get him to stop.

Chapter 23

"James, you sure this is the spot? We've been here for three days and still no ship."

"I may have gotten us here early, it's kinda what I do."

"Why? Why would you want to sit here for all this time? You do realize that accomplishes nothing, except wasting my time."

"I like to get to my shady meetings early to make sure that I will not be double crossed."

"Once again, we are wasting time. All our sensors are off, and we're just floating around wasting time. We can't train because you don't want to miss something happening, but for some reason we can all sleep at the same time. This makes no sense."

"Look, they are supposed to drop off ***your*** new ship in a few minutes. Then we can kill me and start working on the rest of the plan."

"If that's the case, James, when are you going to jump in the med pod? We have to get your DNA altered and change your face."

"I will as soon we get the new ship. I still have to look this way when I die, you know that."

"I know, I know, sweetie. But if I have to wait much longer to fake kill you, I may do it for real. The only thing stopping me is Loki. I think he would be very upset with me, and I can't listen to him whine for the rest of his life."

Loki looked at the two of them, sitting in the pilot's and copilot's chairs. "Loki Yes."

"See," Trish leaned back in her seat. "He would be mad."

Hours passed with nothing happening. The meeting time had long since passed, and Trish was going stir crazy. Finally, the flash of a que-drive saved James from another angry tirade from Trish. The flash was followed shortly by a second flash. James stared at the monitors, waiting to receive images from their camera drones. After a full minute, James and Trish were able to see what was going on. There, in the blackness of open space, sat Trish's new ship, fresh from the factory. James swore he could smell the new ship scent from his seat.

On the monitors, a second ship appeared. This one was a small company ship with the shipyard's name painted in bright blue letters to contrast the light gray ship. James could read the name off the drones, *Apollo Shipworks*. James recognized the name of the company and let out a deep breath as he continued to watch the monitor. James and Trish watched in silence as the three figures leapt from Trish's ship, to the Apollo delivery ship. The space walk only took 30 seconds, and as soon as the last person was aboard, the little ship vanished in a flash of light.

As they approached the new ship, James was shocked at how much larger it was than is own. With the new bounty hunter persona James had, he was able to buy the upgraded weapons package and the ship bristled with weaponry. From the main twin Mass Acceleration Canon's (MAC) running down the spine of the ship, to the thick armor covering the body of the craft, its black paint with gray highlight scheme left no doubt that this was a ship of war. James was certain that soon, the ship would be christened by fire and earn the ship of war title.

"James, she's beautiful. I never thought that I would have a ship like this. Just look at her lines, she almost reminds me of an ancient spear head, or maybe a push knife. She looks like beautiful death. As a matter of fact, I think that will be her name, *Beautiful Death*. It kinda rolls off the tongue."

"I'm not calling the ship Beautiful Death. That is ridiculous, and I'm certainly not calling the ship a *her*."

"Oh, James, I am going to enjoy killing you. As a matter of fact, it might be the highlight of my year."

"You mean *fake* killing me, right?"

Trish smirked, "Ya sweetie, I meant fake killing you."

"Great, let's get this over with."

James started the counter burn and stepped out of the cockpit, heading back to the cargo area. Everything was already packed and ready to be transported to the new ship. After several hours, they had hard docked with *Beautiful Death*. Trish was the first to cross the air lock when an ominous voice seemed to surround them, *Intruder detected. Present authentication codes or be terminated*.

"James, sweetie. Did you spring for the self-defense package as well?"

"Yeeees, it seemed like a good idea at the time," James stammered as he watched the low velocity plasma cannons rise out of the wall. "I think you should give it the codes, Trish."

"You mean give *her* the codes, right?"

"What are you talk-" The ominous voice interrupted James, *You have 30 seconds to input the authentication codes, or you will be terminated.*

"TRISH, GIVE IT THE CODES!"

Trish plastered a huge grin on her face, "You mean give *HER* the codes, right?'

"Yes! Fine! Give her the codes, hurry."

20 seconds to termination.

"What's my ship's name?" Trish asked, still smiling.

"You got to be kidding, I'm not calling the ship that!"

"Oh, I think you will."

15 seconds to termination.

"FINE!!! Would you please give Beautiful Death the codes before *SHE* kills us?"

"That's better." Trish's grin vanished, and she recited the command authentication codes that gave her sole control over the ship's systems.

Authentication codes accepted, Captain. Please procced to the bridge to complete initial set up of the systems.

Trish laughed and turned to James as the plasma cannons receded into the wall, "Now be a dear and start moving our stuff into Beautiful Death." She turned toward the bridge, and James heard her say, "Turn gravity to .3

throughout the ship, please." James felt himself getting lighter, just to the point that he had to be careful not to push off the floor too hard, or he would end up hitting the ceiling. He muttered, "This should be my ship." Loki trotted up to him as soon as James approached the air lock. "Well, buddy, we have work to do. Go get your harness. We are going to be busy."

It took a full day in .3 gravity to move all the contents from James' ship to the Beautiful Death. By the end of the day, James was exhausted. It would not have taken so long, but Trish would not let him just put everything in the cargo hold to be sorted out later. She had a place for everything, and James had to provide the labor with Loki pulling loads alongside him to get all the cargo to its place. Once his ship was stripped of the fabricator, the med pod, all the stores. She swept James' ship with a fine-tooth comb. Once she was satisfied that nothing of value was left except James' EVA suit and mech armor, Trish finally gave the pair a break.

After that, she returned to her ship and activated the med pods. She began printing a James clone. To do this, she had to override about a dozen safety locks put into the med pods programing. The clone would take about a week to print and would be brain dead. As far as she knew, no one had ever accurately transferred someone's complete mind into a clone. With that task done, she met James in the cargo bay. "Okay, James, your clone is being made. When it's done, we will move on to Loki's clone. We still need to plan the fight and select a location for the ship to crash. I also need you to transfer me about two million credits for you and Loki's new suit, as well as operating costs."

James balked at the number, while it was not a lot compared to the money he got from Ski, it was still a tidy sum. "Oh, and James, make sure you pull it out of your bounty hunter's account, by the way, what's his name?"

"His name is Jack Havoc." James said with sense of pride at coming up with such a great name. Trish was on the floor holding her sides, trying to breath between fits of laughter. She was not even trying to hide what she thought about the name. "What... What comic did you lift that gem of a name from?" Trish exhaled deeply, now trying to stop the laughing fit. "Let me guess, you want people to call you Havoc, right?" James stared sheepishly at his feet. "I used a random name generator until I found a name that I thought a bounty hunter would have, and ya, I was going to have people call me Havoc. It sounds cool."

"Dear gods, James, you'll have the coolest name on plaque for your body's ash can in the entire galaxy. Now, come on, *Havoc*, let me give you the official tour and show you to your rack. Stay close, the BD is a much larger ship than your little recon toy."

"I know, I read the manual....this time. Oh, and by the way, what is the *BD*?"

Chapter 24

Trish gave James the grand tour. He knew the ship was larger than his old ship, but he failed to realize how much larger. Of the three decks, the upper deck was his favorite. It housed the observation deck/bar at the bow of the ship. In the middle of the upper deck was the gym, and the last compartment on that level was a Virtual Reality (VR) simulator. The main deck was the middle deck and consisted of six crew cabins at the bow, three on each side of the main corridor. The bridge was in the very center of the ship. The last of the compartments on that level was the Captain's cabin. Trish's cabin was on the right side of the corridor, and on the left side was a briefing/dining area. On the third and lowest level at the bow was a berthing able to house 20 troops, followed by engineering spaces. The last compartment on the lower deck was the med bay. Split by the main corridor on the lower deck on the left side was the med pod and two beds with a door that sealed that part of the med bay off. On the right side of the ship was two exam rooms with separate doors leading to the main corridor. All three decks ended into the cargo area that took up the rear third of the ship. All the decks were connected by a platform lift in the cargo bay as well as a ladder that ran up from engineering section to the observation area. The ship's stub wings mounted to the cargo area and contained the engines as well as weaponry. The ship's main weapon, the MAC's, ran from between the stub wings along the spine of the ship and ended just behind the plazsteel of the observation room.

After the tour, James had to grab his personal gear. It was the last thing in the cargo area that needed to be stowed. As he gathered up his things and headed to the cargo lift, Trish stopped him. "Where do you think you're going?"

"I'm going up to my room. It's on the main level, right?"

"Oh, sweetie, you are in the troop berthing on the lower deck."

"YOU GOT TO BE KIDDING ME TRISH!! This was my ship, I should be allowed a cabin!"

Trish chuckled, "Fine, you big baby. You can have the one closest the bridge port side."

"Thank you," James said through gritted teeth, "and what side is port?"

Trish laughed, "How did you survive without me? It's the left side as you face toward the front of the ship, starboard is the right side. They never change, no matter what direction you are facing. For example, my room is always on the starboard side and yours is always on the port side, got it?"

"Ya I got it, thanks."

"BD cargo lift to the main deck," Trish said to no one. A sultry female voice answered, *Right away, Captain.*

"I get it. BD is short for Beautiful Death, and you made it a female voice because the ship is a girl. Got it."

"Actually, James, BD has a pretty advanced AI and chose to present as a female. After you stow your trash, come to the bridge, and you can meet her. She is a very different ship from your old toy."

"Can't wait." James grabbed his stuff and moved through the bridge to his cabin. As he approached the door, he looked for a handle, button, or some other way to open the door. When he could not find one, he shouted to Trish who was in the bridge setting up her holographic displays. "Trish, how do I open the door?"

"I thought you read the manual."

"I did, but I must have missed the section about opening doors."

"Fine, it's voice activated."

"Great," James muttered, "Ship open door." Nothing happened. "You have got to be kidding me, BD open door." Nothing happened. "Trish, what do I have to say to get the sh- I mean BD- to open my door?"

"Did you try asking nicely?" Trish giggled.

James inhaled deeply, gathering all his patience. "BD, would you *please* open my door?"

"Of course, Havoc."

The sultry voice seemed to surround him, and his door opened. The cabin was twice the size of his old room and even had a desk across from the bed. The cabin shared a bathroom with the cabin forward of it. Loki already made himself at home and was sprawled across the bed. "You know, you can nap somewhere else." Loki 's tail wagged twice. Then, in James' head, he heard, "Loki, no." James laughed and plopped down next to him. Loki rolled to his belly and put his head on James' chest. James was just drifting off to sleep when the voice of BD snapped him awake. *"Havoc, the Captain wants to see you on the bridge."*

"That's going to get old fast." James got to his feet and walked toward the bridge. He stopped when he saw Trish talking to someone who had their back turned to James. "Who is that?" The person turned to face James. He was looking into the electric blue eyes of a teenage girl. The girl was almost as tall as Trish but had not developed into womanhood yet. Her hair was the blackest James had ever seen and hung down just past her shoulders. Her skin was deeply tanned from what little James could see. She was dressed in a blood red spacer jumpsuit with the sleeves pulled up to just below her elbows. James nearly had a cardiac episode when the girl moved so fast she seemed to disappear, only to reappear right in front of him. "Hello, Havoc. I am Beautiful Death, but you can call me BD for the sake of brevity." The voice still had the sultry over tones, but James heard something else in there too. He could not place it until he focused on BD's face. He heard innocence.

"Hell... Hello BD, I did not know you were a real person." James stammered out as Loki circled while sniffing the newcomer.

"I am not a person. I am an artificial intelligence designed to facilitate maintenance, navigation, and the combat needs of the Beautiful Death."

"Um, okay. Nice to meet you. Hey Trish, did you know BD had a body?"

"Of course I did. It was in the manual."

"No, no, it wasn't. I may have missed the door part of the manual, but I am certain I didn't miss the part about an android."

"Did you read the advanced AI manual?"

"Well, I think that we both know the answer to that, don't we?"

Chapter 25

James had never felt so lonely in his life, Loki was on the Beautiful Death with Trish and the very creepy BD android. It had been two weeks since they parted ways, and everyday, James missed them more and more. In his mind, he began to think of Trish as a mean older sister that always looked out for him, but he missed Loki the most. Loki whined and begged the whole time that they were in communication range, so many times, James wanted to go and get Loki, but the next part of the plan was going to be extremely dangerous. The fact that the only two other occupants of his little ship were the clones of himself and Loki did little to settle his nerves. Looking at their inert forms gave him the shivers, so he avoided the cargo area where they were currently housed with their life support systems.

Now that he was in the system where the galaxy would think he died, he had to get to the grisly tasks that he was dreading. James looked at his clone and shivered as he closed the body in the new mech armor he just bought. He turned all the systems in the suit off and grabbed the plasma torch. He heated up the torso to the point of failure. Once he thought that the suit was compromised, he grabbed a piece of plazsteel and thrust it into the weak spot. James instantly regretted not wearing the generic EVA suit and not turning back on the gravity. With bile rising in his throat, he continued to push the plazsteel into the body until the scent of charred flesh and the musty smell of an open torso was overpowering. Blood and shreds of flesh floated in the cargo bay and splashed against or stuck to his coveralls, James continued to push and twist until the plazsteel hit the back of the suit.

James looked around the cargo bay and was horrified at the scene, it looked like a cheap VR sim with the amount of blood and viscera floating and splattered everywhere. James powered on the suit and, with his HUD, watched as the suit tried to save the life of his clone. Chemical cocktail's, electric shocks, and even the chest plate tightened and loosened trying to save the clone, but James had done the job to well. Soon, the suit begin to shut down as clone was unresponsive to any of the treatments. If there had not been such a large hole, the suit would have placed the body in stasis, trying to preserve it for medical assistance.

Now came the hard part, he lifted the clone Loki and placed him in his armored suit, it was odd the clone still had all four natural legs and looked like he was sleeping. James wept as he snapped the last of the fasteners to clone Loki's Helmet. He grabbed a pipe that Trish had fabricated on the BD and raised it over his head. He paused, staring into the face of his best friend in the galaxy, the pipe slipped from his hands and float harmlessly in lazy circles behind him. He could not bring himself to bash the clone, he laughed at himself. "How in the world had I thought I could blow up a planet, I can't even destroy a clone.".

James removed clone Loki's Helmet and smashed it with the pipe until the faceplate was cracked. James was impressed with how much abuse it withstood before the helmet was broken. Once the faceplate was shattered, he placed it back on the clone Loki's head and fastened the clips. James considered the worst part of the plan over, and he settled in for Trish and the BD to arrive and kill him.

. . .

During the two weeks, Trish spent almost every waking moment in the bridge or VR simulator, she planned battle tactics in the ship and with Loki and BD standing in for James. In the beginning, she lost almost every space battle but won almost of the ground engagements. As former special ops, she was used to the ground engagements, but captaining a starship was out of her training depth. She was a fast learner and studied hard. Loki had to remind her when it was time to eat and head to bed. She could tell he was sad, but he trained with her in the VR simulator every day.

When Trish was busy flying and doing the ship training, Loki and BD would go to the cargo bay and play tag or with a ball until Loki would drop out of shear exhaustion. The android seemed to take a liking to Loki and the two only separated when it was time for Loki to go to bed. On rear occasions, Trish would see the pair from the VR simulator. She was amazed how fast the pair had become. In normal gravity, the pair would sprint around the XS-12's or jump over them, they were the only real obstacle in the cargo bay. When the gravity was turned down or off in the cargo bay the pair seemed to become blurs. With no sense up of down, it became common to see one of them standing on the ceiling while the other was on a wall.

One day after an extremely fast game of tag, Loki lay panting on the ground sprawled out trying to cool off. Trish asked BD, "How much faster can you go?"

"Trish," the android seemed to cringe as she said it, but she had been instructed to call Trish that and not Captain, "I was operating at 73% of my maximum speed when interacting with Loki during this session."

"Would you touch all four walls, the ceiling, and floor of the cargo area, then return the VR room at your maximum speed?" It was as if Trish was caught in a vacuum and an over pressure at almost the same time, before she could fully comprehend what was happening, BD came sliding to a stop exactly where she was less than a second before. As BD came to a standing position, Trish noticed that one of BD's feet was missing. "Um, what happened to your foot?"

"I sustained a catastrophic failure of my right load bearing support structure on my first jump. It slowed my progress by 5% throughout the rest of the maneuver. I have already programed the fabricator to make the parts required for the repair. I will be back to optimal performance within 7 hours."

"You lost your foot and still managed to be that fast, sweetie have I got some jobs for you, James is going to hate this," Trish grinned sadistically.

The fight with James quickly approaching, and she still had to make a stop at Hel she was pressed for time, but then again, she thought James seems to like to show up early and wait. She had just enough time to pull into the maintenance hangar and have her ship updated with some of the things James did not think of, as well as a couple of things for her.

James' new assault class mech suit and a new armored suit for Loki showed right before she had to leave the hollowed out asteroid. With all of the alterations to the ship complete and some new very interesting information she left the station at Beautify Death's maximum speed. She should be just a couple hours late to the meeting with James, but it would be worth it. This was going to be fun.

Chapter 26

"She's doing this on purpose, I know she is." James sat in the pilot's seat, grumbling to himself. A flash of a que-drive interrupted his self-pity party. "James Minder, you have a bounty placed on you for the crime of terrorism. You will turn yourself over to me and be taken into custody. If you refuse to comply, I will destroy your ship with you in it, I only need proof of death to collect the bounty, the choice is yours. You have 30 seconds to comply." Trish stated without any signs of familiarly and across all open channels. James began to panic this was not how the scripted the fight was supposed to begin. "Trish would not double cross me, would sh-" James was interrupted by a MAC round skimming across the bow of the little recon ship. That definitely was not in the plan, James thought Trish might actually be trying to kill him.

James pushed the recon ship to the maximum thrust he could handle before blacking out, the Beautiful Death launched a couple of missiles in response, he had to get out of the gravity well and activate his que-drive. "James Minder, this is your last chance, surrender or be destroyed I have two ship killer missiles inbound to your location. Power down your drives, and I will deactivate my missiles." James had to change his heading because one of the ship killers was approaching from the bow of his ship. James opened up his ships self defense menus and set them to automated and weapons free. He flinched as the vibration of the defense turret firing shook him in the co-pilot seat, he looked over to the pilot seat and wondered if he was going to end up like his clone in his mech armor. An explosion rocked the little ship and snapped him back to reality, the defense cannon destroyed the missile approaching from the bow of his ship.

Now he had to worry about the missile coming from the aft, a pair of MAC rounds just missed him off to the port side, James pulled back to starboard and started back to the rouge planet. At this rate, he was never going to get out of its gravity well. The second ship killer exploded off his starboard side way too close for his comfort. "James Minder, this your last warning surrender or perish." Now that was part of the plan, when Trish said parish James was supposed to launch a counter attack and take a MAC in the cargo bay. James quickly snapped his Helmet closed and depressurized the ship.

He turned back into the Beautiful Death and began his attack run, he selected the black and red ship as a hostile target and let the limited AI fire away. He was already down to 70% on his ammunition supply after firing at two ship killers, without the fabricator, he would not be able to make any more and would run out if this kept up to long. Then, a random thought struck him. *Did that crazy woman have the ship painted, is that why she was late?* Instead of the black ship with grey accents, the ship was now black with blood red accents. Running down both sides of the ship in letters at least 30 feet tall, James could clearly see "Beautiful Death" written in the blood red paint.

James forced himself to refocus when the first of his rounds started to impact the static shields of the now terrifying Beautiful Death, not a single round made it through to the hull. James banked back to his port side and gave Trish the best shot at his cargo bay that he could. He continued his turn on the verge of blacking out from the G's, then he braced himself for death or the shock of the MAC round slamming into his little ship. He was not sure which one was going to happen.

The impact was so much worse than expected, his head slammed into the side of his Helmet so hard that he saw stars and became nauseous. James fought to keep the little recon ship from entering a spin that would crush him. His fingers danced across the monitors trying to get the ship to stop spinning and pointed in the right direction. As James steadied the flight path, he realized that he was headed in the wrong direction, in a moment of panic, he rotated the ship while still at full burn.

James' ship disintegrated around him, the added G's to the weakened hull fractured the ship into several pieces. As the cockpit separated from the engines, James' seat was thrown into the ceiling, he felt a sickening crunch and tear at this left shoulder. The last thing he saw before blacking out was his HUD lighting up with more warnings than he had ever seen. His vision dimmed and the stars of the black were replaced with a deep red that seemed to boil before turning into frozen crystals.

Just before Trish was able to activate her que-drive getting out of Hel's gravity well, a large bounty came across the High Value Contracts List (HVCL), a bounty hunters group that Jack Havoc paid to be a member of. The bounty was for James Minder, he was wanted for being the leader of the Free Worlds of the Milky Way (FWMW) terrorist organization, ransoming an entire planet, and detonating two unauthorized munitions of unknown origin. The bounty was 20,000,000 credits alive and 10,000,000 credits dead. There was also a cargo recovery bonus of 5,000,000 credits. Trish placed Jack Havoc's hunters token into the system to show him as actively working the bounty. Before Trish engaged her que-drive and dropped out of the network, five other bounty hunters had put their tokens on the contract as well. Trish smiled, this was going to be fun, and she was going to get paid.

The Beautiful Death appeared in deep space right where she was supposed to be, and only a few hours late. BD notified Trish that she had a contact that matched James' ship. Trish stood at the command station and told BD to broadcast her following message on all open comm channels, "James Minder, you have a bounty placed on you for the crime of terrorism. You will turn yourself over to me and be taken into custody. If you refuse to comply, I will destroy your ship with you in it. I only need proof of death to collect the bounty, the choice is yours. You have 30 seconds to comply." Trish signaled BD to cut off the message. "BD, how close can you put a MAC round to James' ship?"

"Captain, I can peel the paint off without damaging the structure of the vessel." The android stated without a bit of showmanship.

"Well, peel some paint." As soon as the words left her mouth, the 20 pound tungsten slug was propelled at .9c, time of flight of flight was less than a second. Before it fully registered with Trish, the MAC round hit its self-destruct limit and disintegrated well past James' ship. Trish was stunned at just how close the round was to him. "Uh, BD. Did you hit him?"

"No, Captain, I merely grazed the forward section of his bow like you asked."

Trish noticed that the shot had its intended purpose, and James was turning and burning hard. "BD, I want you to listen carefully before you act. Is that understood?"

"Yes, Captain."

"Good, I want you to strip the AI out of two of our missiles, fire one forward and one aft of James' ship. I want to be able to detonate the missiles if they get too close to him, and I also want them to run straight at him, no dodging his point defenses. Do you understand?"
"Yes, Captain."

"Fire the missiles when all requirements are met." Two missiles instantly launched from the stub wings of the Beautiful Death. "BD, open comms just like before."

"Open, Captain."

"James Minder, this is your last chance to surrender or be destroyed. I have two ship killer missiles inbound to your location. Power down your drives, and I will deactivate my missiles." Trish signaled for the comm to be cut. "Come on, James, activate your point defense." Trish pleaded.

"Captain, one of our missiles has been shot down. The second one is closing in fast. It will hit in ten seconds."

"In three seconds, destroy the missile and give him a wide shot with the MAC's."

After the second missile self-destructed, she opened the com channel again. "James Minder, this your last warning: surrender or perish." Hopefully he remembered the code. Trish watched as James' ship turned broadside to her, that was what she was waiting for. "BD, fire a low velocity MAC round through the cargo area of the recon ship, but make sure not to destroy it."

"Yes, Captain." The round left the MAC cannon at .05c time of flight 20 seconds.

"Captain, the recon ship is firing at us, raising the static shields."

"I swear to the gods, James. If you scratch my paint, I will leave you on that rock." Trish watched as the point defense rounds bounced harmlessly off the static shield.

"What did that bring our shields down to?"

"Our shields are still at 100%, Captain. MAC round impact in 3..2..1..impact. The recon ship is still intact but is in an accelerating spin."

"Come on, James, get that thing under control." The recon ship slowed its spin, and for a second, Trish relaxed. As the ship started to accelerate away from the rouge planet, it was supposed to crash into, Trish watched as it flipped 180 degrees under full burn and broke apart.

"Track all the pieces and prep the shuttle." Trish sprinted past her room and leapt off the main deck to the cargo deck. In midair, she yelled, ".3 g's now!" Her leap was extended, and she landed just outside of her new shuttles docking clamp. "BD, vent atmo in the cargo bay and open the doors," she called over the comm.

The small shuttle was out of its docking cradle and burning toward the spreading wreckage of James' ship before Trish had completely thought about what to do next. "BD, are you picking up any life signs?"

"Yes, Captain. I am picking up Havoc's suit transponder. His life signs are weak but there. However, he is no longer in his ship. He is on a basilic trajectory away from the rest of his ship. I am sending the intercept course now."

"BD, I need you to stay tracking all of the pieces of the ship and let me know if they are going to miss the planet we still have to execute the rest of the plan."

"All major parts of the ship will impact the planet; however, they will sustain more damage than we planned."

"That's fine, we'll just have to make it work." Trish plotted the intercept course with Havoc, it would be another 5 minutes before she could start her counter burn. "Well, James, welcome to being dead." She muttered.

Chapter 27

Trish opened the shuttle to space and looked out for Havoc's body floating in the black. She spotted it when it crossed in front of a group of far off stars. When the stars disappeared momentarily, the movement drew her eyes to Havoc's inert form, slowly spinning in the vastness of space. Trish attached her safety lanyard to the shuttle and leapt toward him, her cable was almost paid out when she began to get pelted by slow moving debris. At first, she paid no attention until some of it stuck to her face plate, melted and boiled away leaving a little red smug. Trish was floating through a cloud of Havoc's frozen blood.

Trish started to worry as her Helmet lights lit up Havoc's body, the suit sealed but there were only three limbs. Havoc's left arm was missing from the shoulder down, his suit automatically sealed the wound. Trish had almost used all of her safety line getting Havoc. Once she had him secured, she started to reel the safety line in. Trish had made it halfway back to the shuttle when she saw the flash of a que-drive, followed by a second, then a third. "BD, I need you to come to me now!"

"Captain, I am currently five minutes away at maximum burn. However, with our Delta V's, you would be killed by what I can only describe as the impact."

"What if I'm in the shuttle with the dampeners at max and the shuttle itself at full burn?"

"You would have to give me control of the shuttle in order to give you the best chance at survival." Trish crossed the threshold, "Take control in 2 minutes, I need to strap Havoc in. Make sure you prep the med pod for the full genetic alteration as well as a robotic left arm replacement." Trish set the gravity to .5 and dragged Havoc's body to the copilot seat. She barely got into her own seat when BD came over to the comm's. "Captain, the shuttle will enter full burn in ten seconds. You will most likely pass out due to the G forces involved in the next series of maneuvers."

"Understood, do it." The blackness was almost instant.

BD controlled both ships to meet on an intercept course. At the last possible second, the android spun the Beautiful Death 180 degrees and cut the main engines. The shuttle barely cleared the cargo door before slamming into the VR simulator and crashing to the deck in the .2 gravity of the cargo bay. Trish woke up and started moving her limbs to make sure nothing was broken. "BD, I...I need you to tell me what happened while I was out."

"Captain, I am moving back to our original course. I am still tracking the debris from the recon ship, and there is comm request from one of the three ships approaching from the spin ward of the galactic center and 20 degrees above our horizon."

"Send the comm to me, get our MAC's pointed at them, and raise the static shields. Back burn us to still follow the recon ships orbital impacts. And give me .5 grav in the hangar bay and the med bay."

"Patching them through now, Captain."

“And what kinda name is Beautiful Death, anyways?” An overly raspy voice sounded in Trish’s head.

“Excuse me… you seem to be running your mouth on an open comm. Keep it up, and I will show you what her name means.” Trish used her sweetest voice.

“Well, boys, we’ve got a live one. Tell me, Miss Beautiful Death, do you have a name?”

“I sure do, sweetheart, you can call me Captain Keller. Who am I addressing?”

“Name’s Gus, ***sweetheart***. We’re one of the top bounty hunting groups in this sector of space, so I suggest you take that pretty ship of yours and move along. We have a bounty to claim.”

“Gus, Gus, Gus. I’m afraid you’re a little late. I have already eliminated the bounty and am in the process of retrieving his body, so I suggest you and your friends go collect another contract.”

“I don’t think we need all this formality. What’s your first name? Come on, let’s be friends. What do ya say?” Gus asked with fake politeness.

“I say it looks like you’re trying to cut in on my bounty. I would kindly ask you to take your friends and leave.”

Trish switched comms, “BD, launch four of the combat drones Diamond formation with us at its center 10,000 mile spread now. Charge the MAC’s and make a show of it.”

“Whoa, Whoa, Whoa, and here I thought we were becoming friends.” Gus fainted to be hurt.

Trish made it to the med pod and began stripping Havoc's EVA suit off. As soon she broke seal, his shoulder started to bleed profusely. She quickly stripped the rest of the suit and closed the med pod. With no time to double check BD's medical adjustments, Trish ran through engineering and began to climb the ladder to the bridge.

"Gus, please leave. I have work to do, and I don't need you to be floating around making my job any harder than it has to be."

"Ok, fine, but we will remember this. All we wanted was to share in an easy payday." Trish watched as three flashes appeared on the main screen of the bridge as she entered.

"All three ships have entered their que-drive warp bubble, Captain. Would you like me to recall the combat drones?"

"No, put 'em in orbit around the target planet and face them looking into space. I want to know when Gus and his friends show up again."

"I suggest we power down the static shields and MAC's to avoid destroying a power coil, Captain."

"Where's Loki, is he alright? Oh, and ya power them down."

"Loki is fine, Captain. I have him locked in Havoc's room and set the bed to crash mode. The maneuver to capture the shuttle could have killed him had I not done it. Why did we need to get you and the shuttle back that fast? The other ships in the system just left?"

"They would have attacked and destroyed the shuttle if they were given the chance, and I kinda like living. Anyway, let's get to that planet. We are about to get paid."

Chapter 28

BD placed the Beautiful Death into a geo-synchronous orbit above the crash site of James' recon ship on the rogue planet. "Okay. BD, make sure you keep monitoring the drone feeds. I have a feeling Gus and his friends are going to come back."

"Yes, Captain. I will also ensure that Havoc's genetic modifications are taking effect on schedule as well as his prosthesis being complete."

Trish took a deep breath and closed her helmet, "Well, Loki, here we go." She reached over and closed his helmet. "Loki ready."

"BD, open the cargo bay."

"Yes, Captain."

The shuttle had a slow leak when exposed to vacuum, but Trish had already figured that may have happened when she saw the damage to the shuttle's aft section from the hard landing that had happened 12 hours earlier. Once the cargo bay was open, Trish gently throttled up the shuttle and brought it into open space. Once she was satisfied that the shuttle would make the trip to the planet and back, she started the de-orbit sequence.

The gray-brown planet spread out below them and she started to scan the area that BD had indicated the majority of James' ship had crashed. With the lack of atmosphere, Trish was not worried about the ship catching fire, what she was worried about was the impact destroying all evidence of clone James. She got a hit on the scanner right where BD said it would be. The main fuselage was a crumpled lump of twisted metal lodged in between a boulder and the ground.

Trish landed her battered shuttle within a few feet of the remains of the recon ship, and the pair walked over. The parts of the ship intermingled with the rock strewn landscape. Trish carefully stepped into what once was the cargo bay and saw the fake XS-12 cases smashed into a pile of scrap. She moved through the dangling cables, careful not to touch any of the live wires. Loki walked just forward of Trish, looking at his former home. As they approached the cockpit, Trish called him back, "Loki, remember James' body is going to be up there. It's gonna be okay though. Havoc will be taking care of you from now on."

"Loki James, no."

"What do you mean?"

"Loki James up." He faced the cockpit. "Loki No James."

"How do you know?"

"Loki, smell James up."

"Okay, Loki, here's the deal. We need to tell everyone that James died here. We can't call James by his name anymore. We have to call him Havoc. Do you understand?"

"Loki, yes."

Trish set her suit to record as she passed James's room, she poked her head in, looking around for anything that would help prove that James had died on this lifeless planet. After checking both cabins, she headed into the cockpit, looked at the pilot seat, and grimaced. The James' clone was decapitated by the boulder that ship was wedged under. There was something odd, though. There was a piece of plazsteel sticking out of his torso. Trish looked around to see if she could find where the plazsteel could have come from. There was not a logical place that beam could have come from. "What were you thinking, James? No one's going to buy this."

"Captain, I have three que-drive signatures approaching the planet from the far side. What are your orders?"

"You've got to be kidding me. I need to fix this mess. How much time do I have?"

"Two of the ships are breaking toward the Beautiful Death's combat drones, and the third will be at your location within five minutes. Would you like me to destroy the ship inbound to your location?"

"Warm up the MAC's and set the static shields. Keep the other two off me. I might be able to make this work. Leave the third ship to me."

Trish had to move fast. The pair ran to the shuttle. Trish gathered up all her weapons and sprinted up to a pair of boulders a few hundred feet away from the crash. "Loki, activate your active cloaking, and don't open comms until the shooting starts. Do you understand?" She watched as Loki's form slowly blended into the terrain. "I guess I'll take that as a yes." Trish switched her active cloaking and blended into the rocks as well. She remoted into the shuttle, set to evasive maneuvers, and plotted a rendezvous course with the Beautiful Death.

The shuttle had barely cleared the horizon when a string of auto-cannon fire ripped into it. The shuttle blew apart in a brilliant show of light that showered the landscape in an eerie blue light. "I guess I know what kinda party this is going to be." Trish whispered to herself. She watched as the ship landed in the exact location that her shuttle just left.

Trish sighted down her S&K Plasma Core Rifle (PCR) 23, watching the ship that had just destroyed her ride. This was going to be tricky. It looked like someone was manning the auto cannon, or they were set to an automatic sentry mode. With the ship being slightly larger than James' recon ship, it left a couple of blind spots that she could approach from if she was careful.

Trish cleared her visor so Loki could see her face. She then pressed her visor into Loki's. "Loki, if you can hear me, clear your visor." After a second, Loki's visor cleared. "Good boy. Now, we're going to stay comm silent. Nod your head if you understand." Loki nodded. "Perfect, I'm going to sneak up on that ship and see if I can make it mine. You need to stay here and let me know if you see anyone." Loki shook his head inside his Helmet. "Yes." Loki continued shake his head no. "Loki, you are going to stay here." Loki barked. Trish felt and heard the vibration. "Fine. But if you get hurt, I'm telling Havoc it was your fault."

Trish and Loki slowly moved from cover to cover. Each time, they got closer to the rear entrance of the ship. "They must be waiting to see if they can get space dominance before pushing to the wreck." She muttered to herself. As she was completing the thought, the hatch opened, and she was face to face with three armed men in combat armored

EVA suits. Trish was caught in the open, 20 feet from the ship. Her active cloaking gave just enough time for her to raise her PCR-23 before the men's suits picked out her signature.

The first rounds slammed into the lead bounty hunter and knocked him into his shipmates behind him. Trish did not waste a second. She continued to fire round after round into his chest, trying to burn a hole through his armor. After 30 rounds, the armor failed, and blood begin to pool on the charred armor covering the man's chest. The second the two men recovered, they were drawing weapons on Trish as she barreled straight toward them.

She felt the rounds ping off her chest plate as she closed the distance and drew her mono-blade. Warnings popped up on her HUD as she crossed the threshold of the ship. Her armor was taking damage at an alarming rate. She had to end this quickly. She drove the blade through the weak spot in the neck of the second man while using his body to block the third bounty hunter's shots. She pulled the blade through the neck and partially decapitated him. She continued to charge into the ship with the second man as her shield.

Everything was going well until she cleared the airlock and burst into the largest open space in the ship. The space was twice the size of the cargo area in James' recon ship, and unfortunately for Trish, there were two other men with weapons ready and waiting for her. She shoved the dead man off her blade and swung the blade at the man to her right. He was quick to duck under the swing and she felt impacts on her back, followed by more HUD warnings. Her mech suit was holding, but she would not be able to take much more of this. As she turned to face the rear attacker, she saw a blur of motion as Loki slammed into him at full speed. The other bounty hunters froze at the appearance of a second intruder in their ship.

Trish did not waste a second and threw the mono-blade at the man closest to the cockpit. The blade flew through the ship and embedded itself deep in the bounty hunter's thigh. The weight of the short sword pulled it down through the thigh bone before the man grabbed it and stopped its momentum. She spun and pulled a pistol off her thigh holster, jammed it into the closest man's throat. Trish pulled the trigger until the pistol ran dry. The flexible armor held, but the tissue underneath did not. The man died gasping for air and clutching at his helmet. Trish did not wait to see the result of her handy work. She sprinted to the man next to the cockpit, who was bleeding out. She gripped the handle of the mono-blade and pulled it free, taking the man's leg and fingers with her as she spun on the only remaining bounty hunter.

"Transfer command of this ship to me, and I won't kill you," Trish said overall open comms.

The man, still on his back wrestling with Loki, froze when the mono-blade came to rest less than an inch above his visor. Loki slowly backed off the man and faced away from the pair, watching for anyone to try to sneak up on Trish. "I don't have the codes, I am jus-" Trish slammed the blade through the visor and into his head, stopping him mid-sentence.

"Ok, Loki, I need you to harness up and start pulling the bodies out of the air lock."

"Loki, yes." A small magnetic puck lowered from his back and he walked over to the closest man to the airlock and put the puck on the center of his chest. The puck latched, and Loki easily pulled the man's body out in the low gravity. Trish reloaded her pistol and flung the blood from her blade. "Time to finish the job." She crept through the now silent ship, pistol in one hand and mono blade short sword in the other. She cleared the four empty rooms before reaching the cockpit's sealed doors.

Trish banged on the airlock with the hilt of her blade. "There are two ways this is going to go. The first is me putting a couple of charges on this door and blow it. The second, and the only way you might survive, is you come out, give me the command codes, and I let you walk off the ship to be picked up by your little friends upstairs." Trish waited for a couple of seconds, then began pulling breach charges off her mech suit. The doors slowly opened, revealing a large man with both hands in the air. "Command codes, *now*."

The man in the armored EVA suit gave Trish the codes as she escorted him off her new ship. Once she verified the codes worked, she forced marched him out of the ship. As they passed the bodies of the man's fallen commands, he turned towards Trish. "You didn't have to kill Nat. He was just a kid."

"I'm sure Nat was trying to spare me when he was shooting me. Get off my ship and as far away as you can. I'm arming the auto cannons in 30 seconds." The man turned and ran, as soon as he cleared auto cannons firing arc, he was cut to pieces. Trish flinched when the first rounds went off. She pulled the ships weapons systems on her HUD, "That idiot. He never turned off the sentry mode... Uh, I blame that on him. Oh well, on to the next." She turned and went back to the cockpit. "BD, how's my ship?"

BD placed the Beautiful Death into a geo-synchronous orbit above the crash site of James' recon ship on the rogue planet. "Okay. BD, make sure you keep monitoring the drone feeds. I have a feeling Gus and his friends are going to come back."

"Yes, Captain. I will also ensure that Havoc's genetic modifications are taking effect on schedule as well as his prosthesis being complete." BD depressurized the cargo bay and watched the damaged shuttle leave. Being tied into the ship allowed her to move anywhere with a connection to the ship with no loss of the ship's status. She simultaneously monitored Havoc's condition, queued parts in the fabricator for repairs to the VR simulator, monitored the combat drones, and began clearing the damaged material from the shuttles hard landing.

BD was half an hour into her tasks when two of the combat drones spotted three que-drive flashes in a close orbit on the far side of the planet. BD ran the information through her processors and discovered that two of the ships were making an attack run on the combat drones while the third was headed into the surface of the rogue planet. Within five seconds, she plotted the likely courses of all three ships. She would have to notify the Captain that she was going to have contact with a ship in five minutes and thirty-two seconds. "Captain, I have three que-drive signatures approaching the planet from the far side. What are your orders?"

"You've got to be kidding me. I need to fix this mess. How much time do I have?"

"Two of the ships, Alpha and Bravo, are breaking toward the Beautiful Death's combat drones, and the third, Charlie will be at your location within five minutes. Would you like me to destroy Charlie, the ship inbound to your location?"

"Warm up the MAC's and set the static shields. Keep the Alpha and Bravo off me. I might be able to make this work. Leave Charlie to me."

The Captains comm signal was disconnected. She had left BD with unclear orders that sent the android into a logic loop. Once the loop counter hit its 500th iteration, it triggered a failsafe deep within the ship's processing cores. BD would let the drones be destroyed, but she would protect Trish, and then the ship itself, in that order. BD charged the MAC's and set the static shields to full forward. She ordered two drones to abandon their post at the poles of the planet and join her flanks in a combat spread. She sent the two remaining drones on a reconnaissance mission to gather information on the make and weapons of the ships circling the planet toward her.

No shots had been fired and the Captain did not give her permission to fire first in this engagement, so she was moving into a defensive formation. From the surface, she received a signal that the Captain had remotely operated the shuttle to return to the Beautiful Death. She watched the data streams as the shuttle was destroyed by the

Charlie. Shots had been fired. Her rules of engagement had just changed.

She set the two drones already headed to the ships in orbit to launch all their long range missiles, then make close attack runs to test her adversaries' capabilities. Her predictive algorithms did not fail her when all of the missiles, then the drones themselves, were destroyed while landing a single strike on the static shields. BD was going to have to adopt an aggressive plan if either the Captain or herself were going to continue to exist.

The Beautiful Death launched a full salvo of 16 missiles at the ship she designated as hostel Alpha. A full 30 seconds later, she fired two MAC rounds on an intercept course to the projected path of the ship. The MAC rounds and the missiles would arrive at the same time just beyond the horizon of the planet out of sight. The missiles would self-correct to target while the MAC rounds would self-destruct once they missed their target. She would have to wait another minute before she would see the result of her actions. BD flipped the ship 90 degrees at the full burn the inertia dampeners could handle. With Havoc on board, she was limited to the amount of G's she could subject the ship to. The maneuver had them traveling to the planet's weak magnetic north while her opponents were last seen coming from the east and west.

BD limited herself to a 20,000 mile proximity to the Captain, any further than that, and she would not be above the 95% success rate for intercepting munitions launched from the other two ships. The Beautiful Death's sensors picked up on 16 distinct explosions followed by two self-destruct broadcasts. From the signatures, she could tell that four of the missiles impacted. She did not know if any of them got past the static shields.

Warning alarms lit up her consciousness as the drone on her southern flank erupted into a brief ball of flame. The data showed an unguided munition from below the drone had destroyed it. Her logic circuits dictated that the hostel Bravo must have dropped to the planet's surface and then fired straight up, attacking her drone from a sensor blind spot.

As close to panic as she could feel, BD launched a full spread of recon drones and set static shields to 120%. She drained all the power from life support, knowing Havoc would be safe in the med pod's self-contained life support. Alpha was on its predicted path and she could see that she scored at least one direct hit onto the ship. As a result, Alpha would have a blind spot on its port bow that she could exploit if the opportunity presented itself.

Bravo's position became a known quantity when a pair of missiles flashed into her sensor net from below. Both missiles struck her amidships and nearly penetrated her static shields. BD's android body winced as the blows connected. Her position was unsustainable. She was going to have to move away from the planet to give herself time to react to the threat from below.

She had been in combat for 20 minutes when she got an authenticated message from the Captain on the hostel ship Charlie. "BD, how's my ship?"

Chapter 29

BD passed all the sensor data and her assessment of the enemy positions. Trish began to form a plan, but it all depended how much the former Captain of her newest ship was able to get out before she commandeered it. She checked the comm logs, but the former Captain wiped them. "Well, I guess we'll find out." She sent a plain text comm to the Alpha and Bravo ememy ships, "COMMS DAMAGED, MAIN COMPUTER DAMAGED, WEAPONS SYSTEMS ACTIVE. SEND FRIENDLY COORDINATES AND EMEMY POSITION. I WILL FIRE A FULL SALVO ONCE COORDINATES RECEIVED."

Gus answered over the comms on an open channel, "So I guess you killed Miles too, huh?"

Trish smiled. "Not sure who Miles was, but if he was on this ship when it landed, then... Ya, he's dead. But I have some good news for you. I left all their bodies in a nice plie back on the planet. Why don't you swing by and pick them up? I promise not to shoot you in the back."

"Is that so? That is very kind of you. Someday soon, I'll return the favor." The two bounty hunter ships turned away from the planet and hit full burn out of the gravity well.

"Well, I guess that's that. I got everything I needed from the crash, so I think we're done here, BD. I am going to grant you remote access to this ship. I want us to meet above the north magnetic pole. Then, we are headed back to Hel. I'm going to search this ship to see what I can find."

A few hours later, the Beautiful Death and the bounty hunter ship appeared back in normal space on an intercept vector with Hel. As they approached the large hollowed asteroid, Trish began to make out signs of life. First, she saw the navigation beacons welcoming her back. As they got closer, Trish could see the maintenance domes and, finally, the massive airlocks that housed the star ships. She could hardly believe that less than a day ago, she had left this very station with an untested ship and plans to make a fortune. Now, all she had to do was collect the bounty, sell the bounty hunters ship, and figure out what would come next. Trish was fairly certain that the plan would fool everyone for at least a while. She was brought out of her thoughts when she saw the Beautiful Death disappear into its airlock birthing. She smiled as the ship she was on pulled in to the berthing next to the Beautiful Death. She could not wait to get back to her new home.

Havoc woke with the med pod still closed around him and his body restrained. His body itched all over. He knew that he must have gotten a massive injection of nanites to be this miserable. "BD, can you come release me from the med pod?" Havoc was surprised at how his voice sounded. He thought, "Maybe it's just because I haven't talked in a while." The pod started to open and Havoc squinted as the light hit his face, the straps around his arms and legs released.

Havoc tried to wipe the sleep from his eyes but stopped after the metal of his left hand dug into his face. He yelped and held both hands at arm's length to examine them. Havoc looked at his left hand and arm. They were gone, replaced with an industrial prostatic up to the shoulder. It had five fingers, the palm that moved on command but when he rotated the hand to look at the back, it rotated almost 360 degrees. It only stopped when he shouted, "STOP!" Havoc looked up and saw BD standing in the doorway to the room. "BD, I think I am going to need you to explain what happened. Last thing I remember was.... Hey you almost killed me!"

"Havoc, it would be better if Captain Keller relays the events that took place."

Havoc was checking his body to see if anything else was missing. As he looked at his body from the pod's reflection, he realized that the body he was seeing was not his. This body was well defined. He could see muscles in his remaining arm, chest, and legs. Havoc saw that his skin had taken on a light brown hue as if he had spent his life living outside. Havoc slowly lifted the cloth covering his waist, he blew out the breath he was holding. "Ok, so the rest of me is here. That's good news." BD handed Havoc a newly printed set of clothing, pants, tee shirt with one sleeve missing, socks, underwear, and a set of heavy boots. As Havoc started dressing his body, it seemed out of proportion. His legs and arm felt short, the robotic arm was taking time to adjust to as well.

After what seemed like entirely too long to Havoc, he stood up and walked out of the med bay. He looked to his right to see the cargo area of the Beautiful Death. It looked like some sort of crash had happened. He could see a scrap plie waiting to be fed into the reclaimer, but some of the parts did not look like they came from the Beautiful Death. Havoc walked to the center of the cargo bay and turned to the bow, on the upper deck was the VR simulator room, which had been hit with something large. The damage was in the middle of being repaired, but he could see that it was not a minor accident.

"Hey BD, where are Trish and Loki?"

"The Captain and Loki are currently in the bounty hunter ship she captured during the ground battle for James Minder's remains."

"So, I guess that you want Trish to tell me about that as well, right?

"That is correct, Havoc. I do not have all of the relevant information about the ground battle. However, I do possess all of the information about the space battle above the rogue planet."

"Okay, I have a couple of questions I don't want to have to wait for Trish to answer. First one, did you cut off my arm; and if you didn't, what happened? Second, there were two battles while I was out. Third, are Trish and Loki ok? And my last one, which is more of a two part'er; How long was I out? And, where are we?"

"I did not cut off your arm, it was separated from your body when your ship broke apart after you over stressed the damaged frame. There was only one battle, and it took place in two different locations at the same time. Captain Keller and Loki did not sustain any damage, you were the only casualty we had. You were unconscious for 23 hours and 15 minutes. Our location is currently docked at the Hel station level 5 berthing 12A."

"OK, where is the ship that Trish captured?"

"That ship's location is currently docked at the Hel station level 5 berthing 12B."

"You could have said right next to us."

"Correct, I could have."

Havoc groaned, "Is it safe to open the cargo bay door?"

"Yes, it is. Would you like me to open them?"

"That would be great, thanks."

As Havoc stepped out, he could see Trish walking out of a ship he had never seen before. "This is going to be an interesting story," he mumbled.

Chapter 30

"You've gotta be kidding me, there is no way all that happened in a day."

"It's true. Loki would tell you if it was a lie, and if you had a neural bridge in," Trish said as she led Havoc past the crowded market place. Trish made sure James' neural bridge was destroyed when she saw the state of his head. She also destroyed his comm. Now, they had to travel across the station to get a new neural bridge and comm unit. Havoc never realized how much he had come to rely on his comm until he no longer had it.

Hel was the largest frontier station on this side of known space. Unlike K-19, Hel was never meant to be comfortable. The old independent mining colony had been living on the ragged edge for so long, it seeped into every facet of life on the asteroid. While not dark, the dim lights always made it feel like it was approaching dusk, even though it was closer to noon in Hel's local time.

With projectile weapons outlawed on the station, Havoc and Trish both wore their mono-blades. Trish's short sword hung at her left hip carefully incased within its matte black sheath. Havoc had his black tanto tipped mono-blade (the one that cost Loki his leg) in what Trish told him was a scout carry, in the small of his back parallel with the ground. Trish also made him practice with a dull blade to make sure he did not kill himself.

Loki trotted alongside them, talking to Trish occasionally. He quickly became a celebrity, attracting other people at the market; most of them had never seen a dog before. A crowd formed around the trio. Trish was surprised how men and women appeared from the shops to pet Loki. She filed the information away to be used in the future at some point in time.

Havoc was quickly getting used to his new body and arm. He had not stumbled in the 20 minutes since they had gotten off the train that ran along the inner wall of the asteroid. “Okay, Havoc, we’re here. I’m going to stay with Loki and entertain the crowd. Head to the back of that shop,” Trish pointed at a very average looking shop across the street, “And let Buster know that you’re here for the comms gear. Don’t let him charge you more or insert the neural bridge into your neck. I have some things I want to install on it before we let you back on the net. Remember *Havoc,* you’re a bounty hunter. Act like it.”

Havoc parted the crowd, they seemed to move away from him without even being asked. He could get used to this. Havoc stopped at the doors and took a breath. “Act like a bounty hunter,” he said in a mocking voice, “That was helpful. Thanks, Trish.” He pushed on the door to open it, but it did not move. *That’s odd,* he thought. Havoc gripped the door handle with his mechanical arm and pushed again. He felt the door creak under the pressure, but it still did not open. Just as he was about to put a shoulder into the door, he heard Trish shout out from across the street, “Try pulling, genius!”

Havoc pulled the door handle and walked into the shop, trying not to hang his head. He walked to the back and was standing across the counter from a tiny old man working on a neural bridge. "Hey Buster, I'm here for my comm gear." The man did not seem to notice Havoc standing there, so he cleared his throat and tried again. This time more forcefully. "Excuse me, I am looking for Buster so I can pick up my comm gear." Once again, the little man did not pay him any attention. Havoc felt the anger growing inside him and gripped the counter with both hands. Just as he was about to yell at the man, Havoc felt the metal under his left hand start to give way. The man noticed and well, Havoc concentrated on squeezing the counter. Soon, the metal gave way, causing his fingers to punch through the counter top.

"Was that necessary?" The old man asked.

"So, you can talk."

"It would appear so. Now, I suggest you leave before I no longer let you leave." The old man stated as a matter of fact.

All Havoc could hear was Trish saying, "You're a bounty hunter. Act like it." He was all in. He slowly drew his mono-blade with his right hand and held its point an inch above the damaged area of the counter. "Sorry about the damage. Here, let me fix it." He plunged the knife into the counter and cut a circle around it. Just as the metal began to fall, he grabbed the piece with his left hand and crumbled it into a small cube. "There. Now you have a hole for trash. Come to think of it, I bet if I tried, I could stuff you in there." He slowly replaced the mono-blade. "Now, how about we talk about my comm gear?"

The old man chuckled, "If you were not with Trish, I would have that arm ripped off your body and then use the metal to fix my counter as you watched." He tossed Havoc a small sealed box. "Tell Trish she owes me. Now, get out of my shop before I change my mind. Oh, and do remember that you have to push the door on your way out. I wouldn't want a repeat of your entrance performance."

Havoc exited the shop, still fuming. He stomped across the street, and the small crowd parted as he approached them. "I have the stuff. I think we should go. I may have made a new friend in there." The words barely escaped his clenched jaw. Trish looked him in the eye. After a few tense seconds, she burst out laughing. "Ok, wait here. I'll go take care of this." Before he could stop her, she had entered the shop across the street. "Well, Loki, I think I may have messed up. I guess we'll see how this goes." Loki paid him no attention as he was having too much fun with all the people petting him.

Trish exited the shop a few minutes later. "You went full psycho in there, didn't you?" She called out as she approached. The crowd turned and seemed to really notice him for the first time. Suddenly, they all had somewhere else to be. The crowd dispersed just as fast as they had appeared. Now, it was just Trish with her hands on her hips, looking at him with a smirk on her face. Havoc Held up his hands, "I may have overdone it a little."

"We are going to need to have a talk about what you think being a bounty hunter is. But for now, we have to go and hit a net café. I need to load your neural bridge, and I want a public access terminal to do it."

"Trish, I'm sorry about-"

"Havoc, we'll talk later. Right now, we need to move. I need you connected back to the net. Things are about to start happening, and you need to have comms." Trish began to walk back toward the train terminal, following a path Havoc could not see. After ten minutes, she suddenly stopped and ducked into a shop. As Havoc approached the door, she poked her head out. "Stay out here with Loki. But stay close, I'll need access to your neural link in a few minutes." He decided that sitting on a curb and petting Loki was the best use of his time.

Havoc had no way to tell how long Trish was in there, but he was starting to get hungry when she finally came out. Before he could stand, she walked over, grabbed the back of his head, and shoved his chin to his chest. Havoc felt the click in his skull, and his HUD popped to life. After only a second, a warning appeared in his vision. "Trish, something is wrong. All I see is an error code and a warning saying *Incompatible Hardware*."

"Perfect." She pulled the neural bridge out without ceremony and walked back into the shop. Just a few minutes later, she came out again. Havoc placed his chin on his chest. A few seconds later, a fully functional HUD appeared in his vision. "You're gonna have to set your HUD up on the fly. We have to get back to the ship."

Chapter 31

They made the next train and took a seat next to the doors. “So, are you gonna tell me what’s going on, or am I going to sit here in the dark?” Havoc asked as he set up his HUD. “Well, sweetie, remember the story I just told you about Gus and his friends? Well, they’re here. I don’t know if they know we’re here. But, I’m guessing that we’ll run into trouble before we make it back to the ship,” Trish said over the comms in Havoc’s head. He had not mastered that skill yet, so he asked his questions out loud. “So, what’s the plan? We gonna run or fight?” Havoc could see the annoyance in Trish’s eyes when she answered, “You really need to figure out your comm issue. I have no idea if someone is listening in on at least half of our conversation. And to answer your question, the plan is to get back to the ship as fast as we can, and then come up with a plan.” She opened a comm channel with BD, Havoc, and Loki. “BD, please tell me you have good news for me.”

“I have no new developments as it pertains to the bounty hunters we fought with earlier. However, the man who issued the bounty on James Minder contacted me, and he wants the proof of death. Additionally, he wanted to know about the cargo in the recon ship. I have not replied.”

“Who is the bounty backer?”

“Adam Kusmierski. It appears that he was the individual that provided the evidence to have the warrant issued.”

"Trish, that's not great. I figured he was behind the bounty, but to try to get the proof of death personally means he's close."

"You're jumping to conclusions. BD, can you send us the message?"

"It was a video message. Sending it now." Havoc's vision was blocked by the video that appeared covering his HUD. He quickly shrunk the video and shoved it to the bottom left corner of his vision. He listened as he pulled up the manual for his arm. "Congratulations on such lucrative bounty. I would like to personally meet you and collect the proof of death. I was also notified that you collected evidence of the cargo being destroyed I would like that as well. I will be able to make it to Hel within two weeks to meet and pay the bounty. I look forward to meeting you, Mr. Havoc, and Ms. Keller." The video vanished.

"Great, two weeks. I guess we'll have to figure out what to do with Gus before Ski gets here," Trish paused, "Havoc, did the planet Ski was holding for ransom ever pay up, or did they call his bluff?" Now Havoc paused, "To be honest, I kinda forgot about that. BD, could you run a search of the net and see if the ransom was paid?"

"Havoc, I did not find a source that outright stated that it was paid, but going into their finances, I can see several large transactions that amount to the 5 trillion credits."

"Wow, that was all in their public records."

"Of course not, that was in the secure financial records."

“How did… you know what, I don’t want to know. But now we know that Ski has a lot of money to throw at his problems, and we are one of his problems.” Havoc stated.

“No. James was one of his problems, and if I recall correctly, James is dead.”

“I just hope he stays that way.” At that moment, Havoc’s left hand fell off and landed in the middle aisle of the train. Trish just looked at him, Loki hoped off the seat, picked it up, and brought it back to Havoc. “What are you doing?”

“Would you believe? Reading the manual.”

“And you thought now would be a good time to eject your hand?”

“In my defense, I didn’t think it would actually work.” He said as he snapped the hand back into place, then patted Loki’s head.

“Well, don’t do it again. We’re almost at our stop.”

They sat in silence for the last part of the trip. Havoc was going over the manuals for his new arm, Trish was planning on what was coming next, Loki had his head in Havoc’s lap, slowly wagging his tail as people passed. On the stop, before they were going to get off, the train cars almost completely cleared. The hair on the back of Trish’s neck began to stand on end. “Havoc, Loki, we need to leave *now*.” She stood and started to walk to the rapidly closing doors. Just before she reached the doors, they sealed shut. Trish looked back at Havoc and Loki, who were right behind her.

"Spread out. Havoc, get the back door. Loki, stay in the middle. I'm going to push into the next car and see if I am wrong, but I can feel an ambush coming." Havoc walked to the back door of the car and pressed himself into the back wall next to the door. Hopefully, whoever came in would not see him until it was too late. Loki was standing on his hind legs with his front paws over the seat in front of him, looking from Havoc to the door Trish just exited.

At five minutes before the next stop, both the door Havoc was standing at and the forward doors opened. Two men in light armor EVA suits walked in, both brandishing wickedly curved blades. Both men looked directly at Havoc. The man that came int the back door to the car had a neatly trimmed beard and wore a grey EVA suit. He stepped away from Havoc and into the row of seats opposite him. "Who are you?" He asked while putting his knife between him and Havoc.

Havoc realized that these men had never seen him. He might be able to buy time until Trish got back and was able to help him. "Name's Jam...Jack. What's going on?"

"Well... *Jack.* I happen to be looking for a pirate named Captain Keller. I don't suppose you've heard of her?"

"Nope. Never heard of her. Sorry."

"Well then, care to explain why you're pressed up against the wall like that?"

"Easy. When I woke up, the train car was empty, and my dog was freaking out. I've seen enough vids to know something bad was going to happen." The man in the blue EVA suit that came in to the car by the front door stepped closer to Loki, but stopped when a low rumble emanated from Loki's chest. The man's voice was higher than Havoc expected. "You might want to control your dog before I cut him up." Havoc heard Loki's voice in his head yell, "Loki NOW!" as he leapt at the man's extended knife arm. Before Havoc could process what was happening, Blue Suit was on the ground with Loki attached to his wrist. Loki's right metal paw was raking mercilessly across the man's unprotected face.

While Havoc was watching in awe at Loki's sudden change, the bearded man lunged at him with his blade. Havoc threw both arms up in a hasty defense. A metal on metal clang made Havoc open his eyes. The wickedly curved blade was still pressing into his metal arm. Both men stared at each other for the briefest second before a kick to the chest sent Havoc slamming into the train's windows. He bounced off and barely managed to twist around another knife thrust. Havoc brought down his left arm onto the man's exposed shoulder. He winced as he heard the collarbone snap, even through the armor. The man managed to hold onto the knife but took a step back. Havoc stole a glance and saw the man Loki attacked, rolling on the ground with his face shredded. He could not tell if Loki was hurt because he was covered in blood. Loki was approaching Havoc and the man in the blue suit, his golden eyes ablaze, head lowered as he stalked forward.

Out of the corner of his eye, Havoc caught movement. The man in the blue EVA jumped at him, both feet aimed at his chest. Havoc almost laughed at the irony as the man connected with his chest. Once again, Havoc was thrown into the side of the train car; only this time, there was another man half in half out of the seat kicking up at his face and chest.

The man must not have noticed Loki approaching because he was completely caught off guard when the dog clamped his jaws on the man's throat. Before could Havoc react, he saw the knife the man was holding flash toward Loki. Luckily for the dog, the man could not get his arm to reach Loki because of the seats he was pinned against. Havoc quickly gathered his thoughts and pulled his mono-blade. He plunged the blade into the man's chest and let go as the man suddenly froze. With the man unmoving, Loki let go of his throat and stepped back.

Havoc watched as the man's skin went pale, and he began to cough up blood. Havoc was petrified as he watched the man's eye twitch and then slowly lose shine, somehow, the eyes now looked dull. The screams of the other man in the train car stopped Havoc from breaking down on the spot. He drew in a breath and pulled the mono-blade from the dead man's chest. He nearly dropped it when the body had a violent convulsion before going still again.

Havoc untangled himself from the now still body and stepped into the aisle where the other man was still trying to piece together his face. Loki was at Havoc's side covered in blood, but there was nothing dripping from him, so Havoc figured it might not be his blood after all. The pair walked up to the man, and he suddenly went still. Through his fingers, Havoc could see the tattered remains of an eyelid and the terrified eye beneath.

At that moment, the front door to the train flew open, and Trish ran through blood streaming from a wound on her head, "MOVE, YOU TWO! RUN!" As she sprinted passed the man on the ground, she swung her mono-blade, taking his head and hands off without breaking stride. She was out the back door before the pair could react. "NOW!" she shouted from the other car. The trio sprinted into the next train car.

"Spread out. Havoc, get the back door. Loki, stay in the middle. I'm going to push into the next car and see if I am wrong, but I can feel an ambush coming." Trish said as she walked into the next train car. That one was empty, too. Now she knew something was wrong. She wanted to go back and make sure Havoc and Loki were okay but knew that she needed to press on. She walked casually through the car, checking the seats for anybody or anything. She did not see a single bag or person; her body began to tingle as she approached the front of the car. With only two more cars between her and the front of the train, she knew she would spring the trap soon.

As she entered the next car, there it was; four men and a woman coming down the center of the aisle, checking the seats and overhead bins. They froze when they saw her through the doors. The oldest of the men shouted something, and they all charged. Trish pulled her mono-blade and backed into the car behind her. She knew that she was only one car ahead of Havoc and Loki and was about to warn them when a stunner round flew by her head. The flat black rod flew past and bounced off the back wall where it was now lying at the back of the train car, harmlessly rolling around on the floor. When she looked back at the direction it had come from, she saw two of the men running straight at her while the woman and older man lingered toward the front of the car. She no longer saw the fourth man. Just then, Trish became very aware of the footsteps on the roof of the train. She had to warn Havoc. She had just opened the comm when a stunner glanced off her armored ribs. The electric pulse was greatly reduced from a combination of the glancing blow and the armored EVA suit, but she was still knocked to the floor, and her side was numb. Trish decided to act like the hit was more effective than it really was.

Out of the corner of her eye, she saw someone drop off the roof of the train and land between the two cars. Instead of turning toward her, they turned toward the back of the train. *Well*, she thought, *I guess he knows someone is here now*. She turned her full focus to the threat now walking toward her. The two men closest to her were both brandishing stunners. She knew if she took a direct hit, she would be done for. As soon as they approached her slightly twitching feet, she slammed her armored boot into the kneecap of her nearest attacker.

He fell to the ground, blocking the aisle, and he flayed about clutching his shattered knee. Trish used the opening to push off the bottom of a seat and regain her footing. She still felt slightly off balanced, with her side being numb, but she had been through much worse. Trish held her short mono-blade sword in front of her as she faced the three remaining attackers. "I don't suppose you fine specimens of human beings would be interested in calling this a draw, would you?" she asked in her sweetest voice. The only response was she got was a knife thrown at her face that she blocked easily. "Well then, I guess the gloves are off."

She sprang forward and slashed at the man who was stepping over his injured comrade. He was able to dodge the initial attack, but failed to see the upper cut that landed solidly in his chest. The man collapsed and vomited before Trish grabbed his head and smashed her knee into his face. His inert form fell on top of his injured comrade, where the vomit, blood, and teeth shook together in rhythm with the train on its tracks.

Trish decided to press the attack and hurtled the two men on the ground, only to land with an old fashion slug slinger pointed directly at her head. She managed to move her head just enough for the bullet to cut a furrow from the corner of her eye past her ear. The impact made her dizzy, and she instantly felt nauseous. She choked down the bile that rose in her throat and drove her blade into the ribs of the woman holding the gun. Trish looked her in the eyes as she pulled the mono-blade sideways through her ribs. The woman fell, her entrails spilling into the growing mess on the floor. Trish's world spun, and the darkness crept in from all directions. In the second it took her to steady herself, the older man had disappeared. Trish took a deep breath, turned, and ran toward the rear of the train.

She opened the door to the car Havoc, and Loki were in and saw a man laying on the deck covered in blood. Havoc and Loki were also covered in blood, but both appeared to be uninjured. Trish shouted, "MOVE, YOU TWO! RUN!" As she sprinted past the man on the ground, she swung her mono-blade short sword, taking his head and hands off without breaking stride. She was out the back door before the pair could react. "NOW!" she shouted from the other car. The pair sprinted into the next train car after her.

“What are we running away from?” Havoc called after her.

“None of those men sounded like Gus, which means that he probably has more merc’s waiting for us at our stop. I don’t want to be on the train when it gets to the station.”

“Okay, I guess that makes sense. But uh… How are we getting off the train? It’s moving pretty fast?” The helmet closed over Trish’s head, and Havoc heard her voice over the comm, “Tuck and roll baby!” She launched herself onto the tracks and hit the ground with both feet and knees pressed together. She fell to her side, she rolled a few times, and popped back to her feet. “Well, let’s go!” Havoc heard in his head. Loki was the next to jump. He kept all four legs tucked in until he was done rolling and sliding on the tracks. He stood, “Loki fun.” Havoc forgot to close his helmet as he jumped, and his HUD flashed an override code, and the helmet slammed into place. He was out of position and landed on one foot, trying to run to maintain balance. On the second step, he was tripped up and ended up doing several flips and rolls before skidding to a stop on his face plate.

All Havoc heard was laughter from both Trish and Loki, “I could have been seriously hurt, you two.”

“Ya. Next time, try listening. Maybe then you won’t end up face planting and making a fool of yourself. Just be happy your armor protected you. That could have been a lot worse. Now, come on. We need to get off the track before another train comes around.”

Havoc looked at his surroundings. In the distance, he saw what looked like a small cluster of old mining plants. Trish was already making her way toward them, crossing the large tunnel the train was traveling down.

Chapter 32

The old mining post was one of the dozens that littered the tunnels that crisscrossed Hel, located next to the main rail line that ran through the asteroid. It was once a hub activity, but now it was completely abandoned. At one point in Hel's history, the asteroid started to become unstable due to the excessive mining. The corporations abandoned the asteroid and it sat vacant for years. The local lore stated that the asteroid was rediscovered by a colony ship in distress. After an emergency landing and finding the asteroid still intact, they established a small colony, and it became a trading post for other ships heading into the deep black. Many of the tunnels were still unused. However, most were mapped.

Trish was studying the local tunnel map, trying to find a way to get to the Beautiful Death while staying off the main routes. "Hey, BD, has there been any weird activity in the hangar bay?"

"No, Captain. Nothing inside the hangar. However, there is a large group just outside the main airlock. They are currently attempting to override the air lock security. I have taken control of the security protocols and have ensured they will not be granted entry for at least 100 standard years."

"So that qualifies as weird, I bet they will be bringing cutting equipment in soon. Are you still locked down by the station's AI?" Trish asked

"No, Captain. I broke the lock down less than a minute after it happened."

"Good girl. Now, when they cut their way in, I want you to use the least destructive means to defend yourself until we get there."

"Yes, Captain."

Trish turned to Havoc and Loki, "Look, we are going to have to make our way back to the ship on foot. With the route I have planned, it should take us 12 hours with no real breaks. And that's if we don't run into trouble. Seal your EVA suits. We're going outside."

"Are you crazy? A 12-hour EVA? That'll cut into our reserve air supply. If one thing goes wrong, we will die."

"So.... You better hope nothing goes wrong, sweetie. Now, let's get out of this compound and head to the nearest air lock. It should be lit up on your HUD."

A map with several waypoints popped up on Havoc and Loki's HUD. "Loki, yes."

"Ya, I guess. Let's get this over with." It was a five-mile jog to the air lock, but with the suits pulling oxygen from the atmosphere, their supplies did not dwindle at all. Several times during the trip, the trio's HUD displayed air content warnings. Havoc was silently thankful that Trish made them seal the suits. After three miles of running, Havoc realized that he was not even close to tired. He had become used to his new gate, and even the weight of the suit did not seem to bother him. He watched Loki and Trish for a few minutes and could tell they could keep this pace for hours.

Without warning, Trish cut all comm signals and slid to a stop. She bent down and grabbed Loki's Helmet. From where Havoc stood, it looked like she was talking to Loki. He was tempted to open a comm channel and ask her what was happening, but before he could, she stood up and started walking toward him. Loki half ran half stocked further up the tunnel until Havoc could no longer see him. He was mildly shocked Trish grabbed his Helmet and pressed their faceplates together. "Can you hear me?" Her voice sounded slightly distorted, but he understood her, so he nodded his head.

"You know you can talk, right? And as long as our face plates are touching, I will hear you." Havoc smiled and just nodded his head again. "Funny. For once, you're shutting up. I kinda like it."

"Hey." Havoc protested

"Now you can shut up and listen. I picked up some stray radio signals up ahead. Loki is going to check it out, but I don't want us leaking any signal like they are. So, no comms unless it's an absolute emergency. Got it?" Havoc almost nodded his head but thought better of it. "Got it." Trish separated their Helmets and began creep along the wall that Loki ran down. Havoc followed her example. Trish stopped and glared at him, Havoc just held out his hands at shoulder height and shrugged. Trish pushed their face plates together. "What are you doing?"

"Following you."

"I know you went through some sims about this exact situation. How about you start acting like it." Havoc looked down at his feet, he really did know better. After briefly chastising himself, he made his way to the opposite side of the tunnel. The blackness of the tunnel would have terrified him in his past life. Since he got the neural link, the dark no longer bothered him. His suit augmented his own enhanced vision, and he could now see pretty well in the dark.

Havoc kept one hand on the wall and his other on the hilt of his mono-blade. He was just able to make out Trish as the pair crept along the walls. Havoc turned his passive scanners to the most sensitive settings trying to find the same radio waves Trish picked up earlier. Occasionally, he caught static but nothing that would let him know that someone was out there. Havoc froze, then tensed. Something danced just at the edge of his vision, his audio sensors picked up the faintest of tapping sounds.

Havoc tensed as a shape took form, and the tapping became louder. Before he could properly react, Loki burst from the darkness. With just a whisper of sound, Loki came to a stop in the middle of the tunnel, wagging his tail in his armored suit. Trish kept her eyes down the tunnel as she took a knee and pressed her and Loki's face plates together. Havoc watched as the pair stayed in the tunnel, seemingly unmoving. After a few moments, Trish walked over to him and pressed their face plates together. "OK, there three people about a quarter mile up the tunnel guarding the airlock I want to use to get to the surface of the asteroid. Loki says they had long guns. But, that's okay, I have a plan." Trish laid out her plan when she was finished.

"There is no way you are using Loki as bait!" Havoc shouted into her faceplate.

"First, keep your voice down. Remember, we are not in a vacuum. So, sound will still carry if you are loud enough. Secondly, sweetie, you are more than welcome to take his place. But let me tell you this, I do not think even I could hit Loki moving at full speed in this light. And I guarantee you that I am a better shot than all of them." Havoc tried to interrupt her, but was silenced by her gaze. "And for my last point, these guys are more than likely using old slugger throwers. So, they could hit Loki all day and not get through his armor. Now you may speak."

"Okay, so if we can get shot all day by their weapons, why don't we just walk up and stab them?"

"I'm not worried about their guns, sweetie," she said through gritted teeth, "I am worried about their comms. What good does it do us to do all this sneaking around if, while they are shooting, they tell their boss exactly where we are?"

"Good point. I didn't think about that, and I still don't like this. But I don't have a better idea. Fine, let's go."

The trio moved up as silently as they could manage through the pitch-black tunnel. Once they could see the merc's, they noted Loki's counts were correct. Three men stood in a loose circle, laughing and joking with one another. They had on light armor with long guns. Havoc looked at the small cut out in the tunnel containing the airlock. There was no way they could sneak up on the merc's. With Trish's plan, the men would only have one or two seconds to shoot at Loki as he sprinted by.

Havoc squatted next to Loki and pulled back on his harness. Loki used the harness like sprinter blocks and got ready to run. Havoc was surprised at how much force was Loki was able generate. Trish stacked right behind Loki and tapped both of them. Havoc let go, and Loki shot off like a rocket, barking with his suit's speakers blaring. Trish was just a second behind him, moving like silent death. Havoc stood up and sprinted after the pair.

The three men blocking the airlock turned in unison and watched as a blur sprinted toward them, barking loudly. Not one of them had the time to track, point, or shoot their rifles at the blur that rocketed past them. All three of them were still watching as Loki sprinted past them and continued to run. The first man did not even make a sound as a short sword erupted from his chest. With a quick pull, Trish pulled the blade free, and the merc crumbled to the ground. The two other men had just begun to react to their situation when Trish's blade found its next victim with a downward stroke. The second man dropped next to the first. The third man stepped back and raised his rifle. Just as he opened his comm, a mono- blade slashed into his throat through his neck armor. The rifle clattered to the ground, followed by its owner.

Trish looked at Havoc and saw that he was turning a shade of green. "If you're gonna puke, open your visor and face the other way. I don't need that right now." She bent down and pulled off one of the merc's Helmet. She pulled a data cable from her waist and plugged it into the helmet. The software from her suit stripped the encryption running in the dead merc's helmet. New voices started flowing over her comms. "Nate, Nate, answer me. What is going on down there? Why did your teams' vitals tank?" The comm cut out. "Great. Havoc, we are going to have company real soon. Time to move." Loki sprinted back through the tunnel and carefully side stepped around the growing puddle of gore that was spreading across the ground.

Havoc walked back into the small cut out, wiping his mouth with the back of his hand. "Feel better?" Havoc removed his hand, and Trish stared at the blood leaking from his mouth and nose. "What happened to you? I didn't see you get hit. Plus, you had your faceplate closed."

Havoc smiled, teeth still covered in blood, "Well, you know, when you get done throwing up, ya kinda wanna wipe your mouth off. I may have come in a bit hot with my new arm."

"You've got to be kidding me, you punched yourself in the face?" Trish giggled.

"I thought we had to get going. Something about impending doom or something like that."

"Sometimes you just have to enjoy the little things." Trish said and she began to work the controls to the airlock. "BD, we are going to need a ride. Things are not going well on our end."

Chapter 33

BD answered the comm, “I understand, Captain, but there is a slight problem. The station side airlock is open, and the ship side airlock has been partially breeched. If I override the hangar door and open it to the vacuum of space, I could depressurize a large portion of the asteroid. How do you want me to proceed, Capitan?”

“Can you shut the station side airlock?”

“I can, Captain. However, the men and women in between the airlocks will not be able to escape.”

“BD, I need you to get out of there now. I don’t know how much longer we will remain undiscovered.”

“I understand, opening the hangar doors now. My ETA is 15 minutes.”

“BD, I need you to prep our mech suits. I have a feeling that we may need them before this is over. That Gus is pulling out all the stops.”

BD overrode the safety interlocks and closed the station side airlock. The merc’s inside the airlock chamber began to panic as the hangar depressurized. Only one of them had a complete EVA suit and quickly closed his helmet. When the hangar doors finally opened, all the trapped merc’s were dead except one. The single merc grabbed the laser cutter and continued to cut through the airlock. As the Beautiful Death cleared the hangar door, the merc was able to cut open the airlock and sprint to the exiting ship. BD watched the merc leap through her sensors. She determined that the merc would be able to land on the ship next to an airlock if she did not take any action. This was unexpectable. She must act. She fired her thruster just enough to cause the merc to crash into the ship at a higher rate than he could survive. BD registered the impact on her hull, but did not waste any more computing cycles on the matter. As soon as the ship had the maneuvering room, she accelerated toward her crew.

BD turned all her sensors to their maximum settings and picked up the same two ships that had attacked her at the rogue planet. Both ships had their static shields up and 20 combat drones in an attack formation. BD ran the probabilities and concluded that the ships would not attack the Beautiful Death. The asteroid colony would look at any attack on her as an attack on them while she was in a close orbit to Hel. However, leaving the colony would pose a significant threat. Ramping up her processers, she ran thousands of simulations to discover the optimal course of action. With the best-case scenario, she would only get a 20 percent chance of the ship surviving the encounter without sustaining significate damage. She would have to recalculate the probabilities once the Captain and crew arrived.

BD kept the ship as close to the asteroid as she could and still make good time for the Captain and crew. A comm from the Captain interrupted her while she was physically prepping the mech suits. "BD, we just got to the surface. We've got at least 10 merc's bounding toward us. The group in the tunnels must have gotten a comm out. I don't think the slug throwers we picked up are going to be enough to slow them down for long. Fire a couple of combat drones ahead of you and start softening them up for us, just make sure that you use low kinetic munitions I don't want to piss off the locals."

"Understood, Captain. Launching now. Your mech suits are prepared and in the cargo area. In the local space, I have detected the same two ships that attacked us on the rogue planet and they have launched 20 combat drones. Please advise on how you would like me to handle the situation."

"Start shooting their combat drones. Hopefully, they don't shoot back because if they miss, they will have to deal with Hel. We are going to head back to the airlock so we can change into our mech suits once you drop 'em. Oh, make sure you pack the plasma core rounds. I don't want to be in a protracted gun fight."

"Captain, I am not stopping to pick you up?"

"No. The second you stop, Gus and his cronies will take their shot at you. Just drop the suits at my mark and stay close to the asteroid." Trish sent the coordinates to the airlock they had come out of earlier. "I'm pretty sure we will lose comms once we get down into the tunnels. Whatever happens, keep my ship safe."

"Yes, Captain."

The craggy surface of Hel offered a multitude of hidden paths and firing lanes. The trio's problem was that they had no idea how to traverse the labyrinth. Trish instructed Havoc and Loki to follow their digital bread crumbs back to the airlock while she brought up the back of the formation. They had five minutes before the combat drones were overhead and another seven after that before the Beautiful Death dropped their mech suits.

Shots shattering on the gray rocky wall of the crevice they were traveling down snapped Trish's focus to the rear camera in her armored EVA suit. Her HUD showed four merc's bounding over each other's positions. Trish recognized the maneuver from her days in the service. "Great," she muttered to herself. They were no longer fighting the local ruffians, these merc's had formal training. "Havoc, Loki, move faster. We need to control the airlock, or this is going to get bad." They were barely able to run at their full speed due to narrowness and the random twists that seemed to go forever.

Havoc was doing his best to keep up with Loki, but had to tell Loki several times to slow down. Havoc laughed to himself, *at least I figured out how not to fly off this stupid rock.* His suit's automated attitude nozzles pulsed and he was forced back to the surface of the crevice. His HUD showed that they were almost to the airlock. The countdown timer showed just under four minutes until the drones arrived. "Loki, two front," the gravelly voice warned. Havoc raised his slug thrower and charged around the corner.

Havoc felt the rounds impact his armor, then saw the warnings on his HUD: *SUIT INTEGRITY BREECHED DEPLOYING SEALING COMPOUND*. The world spun as impact twisted him around. Time slowed, and he saw the flash of Trish's rifle but heard no sound. He twisted back toward the enemy and saw the rounds impacting against the armor, but not penetrating. His gaze was drawn to their weapons as they tracked toward Trish, plasma core rifles. *That's why they breeched my suit so easily.*

The combat cocktail threatened to crush him for an instant, then he was back in control. As if guided by someone else, he drew his mono blade from the small of his back and lunged forward. The blade entered the merc under his left armpit and nearly took off his arm. Blood and sealant created a red and yellow ice cloud. Havoc threw the blade, and it harmlessly bounced off the second merc's Helmet. He froze as the rifle was turned on him. A black and orange missile collided with the man as he pulled the trigger. The shots missed wide, and Havoc sprang back into action. He jumped and grabbed the knife while it spun on its arc out of the crevice. As his attitude nozzles kicked in, he was driven on top of the merc Loki had just crashed into.

Loki was using his prosthetic to dig at the merc's face plate. The metal alloy claws dug deep furrows into the face plate, but did not breech the plazsteel. The man batted Loki aside just in time to have Havoc dive mono-blade first into his chest. With no time to move, the blade sank to the hilt in the man's breast plate. The merc slammed his rifle into Havoc's helmet, but Havoc was able to hold on to the blade as he was thrown off the fallen man. The mono-blade cut a path of ruin across the man's chest as Havoc flew off. The suit was unable to seal the damage, blood, and sealant mangled in a cloud of death. Havoc looked back to see Trish driving her short sword through the first man's neck, ending that round of the fight.

"Time for the butcher's bill. How bad are you hurt?" Havoc paused, taking stock of himself. As he moved his body to check for pain, he realized he could not feel his left hand. He held up his mechanical stump. "You have got to be kidding me, I just got this." He said, waving the stump for all to see. "Loki, good."

"What about you, Trish? Ya good?"

Trish slumped against the rock wall. "I've been better. I got a few hours before I have to hit the med pods before I'm in real trouble." She moved her hand off her right side, and Havoc saw the suit's sealant closing a hole that went completely through her suit and body. He began to move to her side. She held up her hand and stopped him, "Grab their rifles. We may need them before we get the mech suits."

Havoc gathered up the fallen mercs' weapons and gently handed Trish a plasma core rifle and pistol, keeping a pair for himself. She pushed off the wall with a grunt and stood up straight. "We need to leave, now. Loki on point. Good job alerting us, by the way. Havoc, you're on drag, and I'll be in the middle. Loki, don't get too far ahead of us. And Havoc, don't crowd me. Keep 20 feet of separation. Now, let's move." She faltered as she took her first step, but regained her footing and kept moving.

Their combat drones were now in the overhead, giving them a real time site picture of what was ahead in the trench. "Mech suits inbound." Havoc and Trish snapped their heads up in time to see the Beautiful Death flash dangerously close overhead. Havoc noted how strange it was that a ship that size, at that speed, did not make a sound as it sped by. The trio watched as two objects fell from the ship and disappeared over the lip of the crevice. Havoc smiled when he felt the impact in his boots. He knew they were close once that happened. The combat drones showed a clear path to the airlock. The trio ran as quickly as Trish could move.

Chapter 34

BD watched as the mech suits dropped from the ship. Even in the brief instant, she was overhead, she could see that the Captain and crew were in trouble. The Captain's vitals were creeping toward the red, Havoc had been injected with a combat cocktail, and he had limited use of his left arm. Loki was the only uninjured member of the group. Her processers began to heat up, she ramped up her cycles to her maximum limits. The plan that had the highest probability of working also had a high number of unpredictable variables, but she was unsure if the Captain would agree to it.

"Captain, I have a plan, but it has some risk."

"BD, go ahead. Give it to me. I am starting to run low on options and need an exfil now."

"On the next orbit around the planet, I will drop a third combat drone into your overhead. I need you and Havoc to change into your mech suits as fast as possible. You have enemies approaching from all angles, your armored EVA suits will not be able to stand up to the amount of fire you are about to receive. Loki will have to mount one of the combat drones in the overhead now. I believe I can get him out of there before the fire becomes accurate."

Havoc interrupted, "If that's the case, why don't we all just grab the drones now and get out of here?"

"The fire that you two will receive will compromise your suits before you will be able to grab on. Loki, on the other hand, will be able activate his active cameo and move much faster than the two of you. Once he gets up to speed, I will have a combat drone drop to 20 feet off the surface of the asteroid. He will have to leap on, then I will pick him up on the following orbit. Neither you or the Captain can pull off this maneuver, so I have made an alternative plan."

Trish gritted her teeth, "When does that happen?"

"Loki should land on the combat drone in 30 seconds. He has already started his sprint."

Havoc looked around for Loki and was surprised to only see a ghost image in his HUD as Loki ran faster than he had ever seen. Havoc stared as the ghost leapt into the air. He nearly screamed into the comms when he saw plasma core rounds streak past the combat drone that was coming up from behind Loki. The pair connected, Havoc Held his breath as Loki started to slide toward the back of the drone. Loki's metal claw dug into the drone's armor. He almost slid off the back before he came to a stop. "Loki, okay." Havoc swore that the gravelly voice sounded winded; something he had never heard before. The drone sped up and vanished over the horizon.

Trish and Havoc ran along the trench until they saw the mech suit containers lodged in the ground next to the airlock. "Trish, how are we getting them out of there?"

"The containers are airtight. Once you get inside, seal the hatch, then open the air canister. This will let you take off your EVA suit and get in the mech. Just remember, it's going to be tight and the bad guys will be here any minute. We need to hurry."

Havoc headed to the container marked as his on his HUD. He looked back to see Trish climb on top of the container and then disappear inside. "Guess that answered that question," Havoc muttered to himself. He jumped on top and found a small interface glowing in the corner. He performed the digital handshake, and the hatch in the center popped open. Havoc lowered himself down and climbed over his mech suit. This one was much larger than his first suit and was unmistakably a battle mech; a small battle mech, but larger than Trish's. The container was cramped and Havoc barely had room to move around to find the air canister.

Once he opened the canister, it dumped its entire contents in the container. His EVA suit notified him that the air was safe to breath. As soon as Havoc opened the suit and disconnected his neural link, the world plunged into pure darkness. He felt around the container, found a light source, and turned it on. An eerie red light dimly lit up the space. Havoc shivered as he stepped out of his EVA suit. The temperature was in the negatives and falling fast. He quickly stripped out of his jumpsuit and stored it in a cargo compartment in the mech suit's leg. He began to shake uncontrollably as he climbed in the mech suit completely naked. Trish had warned him about this style of suit and how its plumbing hooked up to *his* plumbing, but it was still a shock when the connections were made.

Havoc closed the suit and felt the posts slide into the back of his neck. His HUD came alive with initial set up options, and he began working on setting everything up. *So far, this is just like the sim*, he thought. The set up only took seconds, and as he was adjusting the suit's power management system, Trish came over the comms, "What is taking you so long? We have company, and I can't do this by myself." Havoc blazed through his menu's, looking for the release icon. "Trish, I can't find the way out of here." Trish yelled over the comms, "You are in a walking tank, I don't care if you punch your way out. Just get out here now!"

Havoc used the mech strength to pry open the container, then he ripped the mounting bracket off the floor and took it with him as he exited. Outside was chaos, he saw Trish's pink armor dancing through armored EVA suits and smashing everyone in sight. He froze as he watched her kick a man so hard, he achieved escape velocity and was sent into orbit. A bright red warning flashed on his HUD, alerting him that he was being targeted. He engaged his auto-cannon and a small multi-barreled rotating gun locked into place above his right shoulder. Before he could find the target, the cannon shook the mech suit. Havoc thought something was wrong, he could see the rounds impact but never heard the gunshots.

He was brought back to his senses when an explosion threw him to the ground. The suit took superficial damage as it slid along the rocky surface. "Get up and start shooting back!" Trish shouted. "What do you think I'm doing?" Havoc got to his feet and took cover behind a rocky out cropping, his auto-cannon sending out a stream of plasma core rounds. He pulled his rifle off his chest and crosshairs appeared in his HUD. He stepped out of his cover and was immediately hit with multiple rounds. He flinched, but recovered when he saw that his armor was still in the green.

Trish was shooting now that she cleared the mercs that tried to swarm her. She was moving from cover to cover, firing, and then repeating the process. Havoc watched as she called in the remaining combat drone to do a strafing run on the entrenched enemy. "Havoc, start shooting, or I am going to start shooting you!" He snapped up his rifle and started firing. He tried to emulate Trish, but could not match her speed and grace under fire. As a team, they started to push back the enemy. His auto-cannon began firing behind him. "Trish, we have company behind us."

"I know." BD'S comm reconnected Trish, she switched comm's to include the Beautiful Death. "BD, where's that other combat drone?"

"It's not coming, Captain. They mined the surface, and I had to take a higher orbit. Loki is aboard with no injuries, but we are outnumbered. I will return to the surface when I can safely do so."

"Don't worry about us. I just need to know if the hangar doors are still open."

"They are, Captain."

"Thanks, talk soon." Trish switched back to the local comm.

"Havoc, looks like we are running to the hangar."

"This feels very familiar; us running to a hangar and people trying to kill us, you know, almost like déjà vu." "Welcome to being a bounty hunter."

"Funny, but I have a question. Why are we going to the hangar?"

"To get my other ship, sweetie." Havoc could hear the pain in her voice this time.

"Loki should land on the combat drone in 30 seconds. He has already started his sprint," BD spoke into the comm as she began rigging the cargo bay to receive Loki, then the Captain and Havoc. She had the fabricator make 20 cubic feet of aerogel to cushion their impact. She placed the aerogel padding along the wall where she planned for the impact to occur, her preparations were interrupted when her sensors picked up multiple anomalies in her flight path. BD was on her third orbit and had followed the same track each time. Her processers concluded that the two spacecraft had set some sort of trap while she was on the far side of the asteroid. The most likely mechanism for the trap was a shape charge anti-ship mine.

BD had to change her orbital track, she then updated with the rendezvous with the battle drone carrying Loki. She calculated that the impact would be increased by 10% and happen in 45 seconds. She did not have time to make more aerogel; instead, she quickly rearranged the aerogel she had and opened the cargo bay doors. Through her sensors on the ship, she was able to track the battle drone and adjust both of their vectors. Ten seconds before the impact: "Loki, jump straight up now!" Without a second of hesitation, Loki jumped as high as he could; the battle drone sped up to its max acceleration, and Loki watched as it disappeared in the inky blackness.

Loki watched as the top of the Beautiful Death began to grow extremely fast in his vision. He activated his thrusters in an attempt to slow down, but he could tell he was going too fast. "Loki scared." The ship grew larger. "It will be all right, Loki. Just keep firing your thrusters and bend your legs before you land, you will be fine. I promise." BD took a cycle to wonder why she had just made a promise that she may not be able to keep. Loki listened and bent his legs, preparing for the impact, at the last second, BD rolled the ship so fast that it would have killed any organics on board. Loki spasmed as his landing spot disappeared and suddenly, he was falling through the cargo bay. He felt the gravity pull on his body, but he was still flying toward the far wall. Just before impact, he slowed dramatically as he entered the aerogel. He hit the wall hard, but did not hurt himself. He lightly bounced off the wall and then hung suspended in the aerogel. "Loki, okay. Loki Help."

"No, buddy," BD had heard Havoc call him that, and Loki always responded well. "I need you to stay in there. I will be maneuvering the ship in a way that will hurt you if I let you out. I will get you out as soon as I can, I promise." There it was again. Another promise. "Loki, okay." Loki wiggled around until he could get into a comfortable position, then curled into a ball and went to sleep floating in the aerogel.

With that taken care of, BD, dedicated all of her processing power to the enemy vessels, their combat drones, and the suspected mines. She had five total combat drones and 10 recon drones at her disposal. Her processors began to heat up as she overclocked them to run more simulations. After 20 seconds, she had her plan.

The Beautiful Death was coming back into comm range with the Captain and Havoc, so she pinged the mech suits she dropped. The Captain's suit came back with a positive ping, but Havoc's did not. The Captain's suit was sending back her vitals, and they were trending toward being combat ineffective. "Captain, I see that you are badly injured. Please advise."

"I will make it, what's your status?"

"Captain, I am going to have to make a hard burn to give myself room to fight. The ships are laying shape mines, and I will not be able to continue my close orbit much longer."

"Do what you have to. Just keep me in the loop"

The comm was cut as the Beautiful Death disappeared over the horizon. BD ran the simulations and was sure that if she launched a comm repeater to maintain the signal, it would be destroyed by the enemy. As she approached the far side of the asteroid, she flipped the ship so the main engines were facing the asteroid. She made sure that Loki was still secure in the aerogel, wedged her android body into a corner, and engaged the drive to the maximum thrust that Loki could take without seriously injuring him.

The inertia dampeners were over whelmed almost instantly, and Loki began to sink toward the back of the ship as the G's piled on. The Beautiful Death's frame groaned in protest, and BD's body was pressed into the wall to the point the wall started to dent in the shape of her body. "Loki hurrrr," Loki's gravelly voice trailed off. BD checked his vitals and saw that he had passed out due to the excessive G load. She would have to maintain that acceleration for another 20 seconds if she wanted to engage into the upcoming space battle at the optimal speed. In another ten seconds, she would be far enough away for the enemy to see her over Hel. Then, the fight would begin.

Once the Beautiful Death reached its desired velocity, BD changed vectors to start her attack orbit. The Captain's comm reconnected, "BD, where's that other combat drone?"

"It's not coming, Captain. They mined the surface and I had to take a higher orbit. Loki is aboard with no injuries, but we are outnumbered. I will return to the surface when I can safely do so."

"Don't worry about us. I just need to know if the hangar doors are still open."

"They are, Captain."

"Thanks, talk soon"

BD noticed the Captain's vitals continued to trend toward combat ineffective. She filed that information away, there was nothing she could now. She needed to focus her processing power on the upcoming battle. She checked on Loki and saw that he was uninjured. After a quick self-diagnostic check that showed all systems in the green, she burned toward the two ships that had so far proven to be formidable opponents.

Chapter 35

"Trish, there's no way this Gus guy has this much pull. We have to be talking over a hundred people that that we have encountered so far." In his past life, he would have already been winded by the amount of running and gunning; but with the gene edits and combat stems, he had never felt better.

"Agreed. These aren't regular mercs, Havoc. I think there is another player involved." Trish ducked around a corner at a plasma core round smashed in the canyon where she had just been.

"These guys are too well disciplined and are fighting as a team. I don't know many merc teams that can do that."

Havoc fired his rifle as he cleared a bend, killing a pair of opponents in lightly armored EVA suits. "I'm not sure about that, Trish, this seems pretty easy. I mean, those last guys were not even in heavy armor."

"Two things: the first is, I am pretty sure that we broke through the front lines of troops that they had. And the second, and this is a big one, sweetie, you are in a fully up armored mech suit. Think about how many times you would have been killed already if you hadn't been wearing that thing. I count at least 20 shots that would have taken your head off. They were not prepared for two mechs. That's why it seems easy," She said while running past him and clearing the next bend. Havoc thought she was not moving like normal, *almost sluggish.*

"So, who do you think is after us? And are you okay?"

"It could be a number of people. I mean, you are the leader of the FWMW." She did not answer the second part of his question.

"No, James Minder was the leader of the FWMW, and that was not even true. The FWMW put a hit out on me... him.. whoever. No one knows that we are the same person."

Two shots flew past his head as Havoc leapt out over the ledge of the canyon. His auto cannon ripped off another string of rounds. He was down to 50% of his ammunition reserves, and a few parts of his armor where starting to creep into the yellow.

"True. That means this has to be about the XS-12's then," She coughed, "But that's a problem for another time. We are almost to the hangar. Get back in the present. I want you to take lead when we get in the hangar. I am trans....transferring the Captain codes to you. Get us into open space and work the plan BD comes up with," Trish said through grinding teeth.

"Trish, are you ok?"

"No, I need to get to a med pod now. The combat cocktail and my suit will not be able to keep me going for too much longer." She planted her back against the hangar door inches away from the opening. "Remember: in this, speed and violence are our friends. I'll be right behind you." She tossed a pair of frag grenades into the hangar. "We go in 3." Havoc stepped in front of her. "2." He moved his rifle into his shoulder. "1." Two explosions ripped through the hangar, and in their wake, Havoc and his auto cannon erupted into the smoke and carnage.

Havoc almost panicked when we saw just how many troops were waiting for them. However, most of them were still recovering from the pair of frags. His auto cannon spun to life and began to cut the enemy down. All at once, the troops seemed to come to grips with reality and started firing back. Havoc started to pick targets out and fire back with his rifle. More of his armor moved into the yellow as the weapons fire impacted his suit. The defenders hid behind crates of space ship parts, pallets of cargo, and servicing equipment, only exposing their rifles by firing using only their gun cameras. Havoc almost stepped back behind the safety of the hangar door when he heard Trish yell over the comm, "Assault through the ambush, MOVE!" Havoc found his courage and stepped forward, bullets started to crack his armor as he walked past the nose of their recently acquired spaceship.

Havoc felt something heavy hit his back. No new alerts showed up on his HUD, so he ignored it and pushed on. He used the fuselage as cover and fought his way to the rear cargo hatch. With each step, he felt as if he was pulling a grav-car behind him. Havoc finally made it to the cargo door just under the port stub-wing. The fire started to slacken as the shooting angles were reduced by his close proximity to the ship.

"Trish, I'll hold here. Open the doors." Havoc received no response. "Trish, get the door." Nothing. Havoc set his auto cannon to sentry mode and then started cycling through his cameras. None of his cameras showed anything. He switched to the drone footage, and his heart sank. The camera angle was slightly off the nose of the ship. Where he expected to see his mech suit, he saw a pink mech suit attached to the back of his suit. Havoc tried to turn but the way Trish attached herself, he could never get a good look at her.

"TRISH!" No response. Panic began to creep its way into his mind, his judgment started to cloud. A small blinking blue icon stopped his spiral. He selected the icon, and Trish's status popped up into his HUD. He now had complete control of her systems and could see all her vitals. He was both relieved and terrified when he saw that she was still alive, but that her suit had put her into stasis due to her blood loss. He needed to get her into a med pod immediately.

Havoc used the Captain codes to open the ship up. As soon as he entered the cargo area, he used the codes to start up the engines. He set the ship's point defense cannons to sentry mode with low velocity rounds. Now that he had atmosphere in the ship, he heard and felt the concussion of the weapons firing. "Alright, we are safe for now. Need to get Trish to the med pod." He looked around for the med pod and found the control panel right where the user manual said it would be. He walked to the front of the cargo area before he noticed that he would not easily fit through the passage way.

Havoc took a breath and unsealed his mech suit. He felt the prongs slide out of his neck as his HUD disappeared. He cringed as the suit's plumbing disconnected from *his* own plumbing. Havoc walked over to the med pod tucked into the wall and opened it up. He then set it to run a diagnostic on Trish and then repair the damage it found. With that task completed, he moved back over to Trish's suit and hit the emergency eject on the control panel on the back of her suit. Nothing happened. He tried again with the same results. Frustrated, he tried to pull up a menu on his HUD. After a few seconds, he realized that his neural bridge was not in, and he had no HUD. He reached into his mech suit and felt around for the bridge. After a short search, he found it and quickly inserted it into the back of his neck.

His HUD sprang up in his vision, and he was able to see that he had to enter the control codes for Trish's suit. As the back of her suit opened up, blood started pouring from the seams. Havoc was horrified to see Trish's nude form covered in blood. The hole in her side started actively bleeding again, and her body began to fall to the floor in the 1g that the ship was set to. He quickly caught her and carried her to the med pod. Havoc shut the pod and set it to repair the damage as quickly as possible. She would have another couple scars to add to her collection, but he did not think that she would mind.

The ship shook violently, and Havoc's HUD lit up. The ship was taking damage from some sort of homing missile he had never seen before. He switched the point defense cannons back to point defense using his HUD while he ran to the bridge of the ship. As he ran, he noticed that this ship looked very much like his recon ship, just bigger. He slowed when he came up to the bridge doors and noticed marks that looked a lot like plasma core rounds had slammed into the walls and door. "What did you do to get this ship, Trish?"

As he walked through the opening bridge doors, an urgent comm request came through to both his HUD and the bridge comm system. Havoc keyed the accept icon, and BD's image appeared in front of him. "Havoc, we are in trouble. I need you to make a full burn to these coordinates. We will rendezvous and enter the warp bubble together." Distortion started to overtake her image shortly before she disappeared.

Missile threats flooded BD's sensors. Gus had laid a trap, and BD had just flown into it. BD ran the calculations. If the Captain and Havoc were aboard, she would not have been able to escape without taking substantial damage. She simultaneously calculated the probabilities of escape with Loki on board. She was certain that the ship and Loki would complete this phase of the battle without any permanent damage.

BD checked Loki's vitals before she started the high g maneuvers to establish a baseline. Once satisfied, she started to spin around the long axis of the ship. Her point defense and counter-missiles flew forward to intercept the incoming threats. BD paused for nanoseconds to admire the mathematical beauty of the ship sending helical strings of auto cannon rounds and counter-missiles. Her processers drew similarities with a spiral galaxy and the Beautiful Death at its center. A cone of destruction erupted surrounding her. She was bathed in radiation and fire as she continued to close with the two adversaries.

Gus's ships continued to pump out missiles and high velocity auto-cannon rounds, hoping to get through the Beautiful Death's defenses. Due to the maneuvers that the

Beautiful Death was completing, it was clear that the ship was un-manned. Gus and his band of mercs changed their tactics. He ordered all ships to cease fire of conventional weapons, then had his team begin to fire the electromagnetic (EM) munitions. Gus also maxed out his own EM jamming fields and EM directional weapons.

For the first stages of the battle, BD had an unobstructed view of the destruction that surrounded her. As soon as the EM munitions detonated, her view became almost non-existent. As Gus began mixing conventional warheads in his spreads, her envelope of safety became much smaller. Rounds and missiles began to detonate on her static shields. BD knew she was blind, but Gus could only use optical trackers to see her. She checked Loki one more time and then oriented the ship 90 degrees solar south while still traveling along her original path.

She programed two decoys and launched them along her current vectors. When they achieved enough separation, she activated them. One continued along her path, cloning her EM signature. The second cloned her EM signature, but it was programed to replicate a full burn 90 degrees solar North from her original flight path. She pushed the engines to 110% of their rating for ten seconds. Any longer, and, Loki would be pushed out of the aerogel and smashed into the back wall of the cargo bay.

After the ten seconds, she shut down all systems to include life support, and only ran a single passive camera. If she could feel nervous, BD was sure she would be as she cleared the worse of the EM bubble. Her single camera showed the battle still raging without her. She watched as the solar north drone was located by Gus when it cleared the EM bubble and destroyed. Even the explosion was what the enemy would expect to see if the Beautiful Death was destroyed. Without the fire power of the actual Beautiful Death, the clone that stayed on her original vector was destroyed in short order. Once again, it replicated an explosion that the enemy would expect to see.

BD traveled from the bridge, past the Captain's quarters and dining hall, and jumped into the Cargo bay. She calculated her jump to ensure that she impacted the aerogel as close to Loki as possible. With no gravity or lights, she had to be very careful trying to dig him out. Once she got him free, she removed a panel from her left arm and pulled out a universal comm cable. She felt around until she found the port on Loki's suit.

"Loki, I need your help moving some missiles and mines. I also need you to keep your neural link on receive only. If you understand, bark. I will be able to feel the plazsteel of your face plate vibrate through my fingers as long as I am touching it." She felt the vibrations. "Good boy, I will be working in the cargo bay to throw out the missiles and mines by hand. I need you to crawl into the magazine and push the ordnance down to me. Once I open the cargo doors, I don't want any light coming from the ship. So make sure that you only turn it on when you need it." BD felt the vibration on her fingers once again. "Let's get started."

Loki crept along the ceiling of the cargo bay to reach the hatch of the magazine. He used the manual override lever to open the cramped space. Once inside, he had to turn on his suit lights and make his way to the missile racks. It was slow going at first, but once he figured out how to remove the safeties with his paws, he was able to start moving the ordnance by bumping it out of the cradles and then hooking them to his harness. After the first few, he was able to move two or three a minute.

BD watched as the missiles and mines were pushed toward the floor. At the rate Loki was sending them to her, she was able to easily catch them as they slowly fell towards her. She then walked over to the open cargo bay doors and threw them in a precise trajectory that would leave them floating between the Beautiful Death and Gus' ships. She would walk back and catch the next missile or mine and then repeat the process. After an hour, she jumped up to the magazine hatch and plugged into Loki's suit as he was prepped to drop the next mine. "We have enough out there. They are still looking for us and they are now in position for the next part of my plan. Come out of there, and let's began the startup sequence for the ship."

BD floated to the floor with the last mine and threw it to the planned area of space. She turned to Loki and paused a random course of action appeared in her processers. BD turned to Loki. "Race?" Loki bolted for the bridge before she was able to finish the sentence, almost breaking her comm cord. In the pitch black, she only got lidar return as he jumped to the main deck, pushed off the ceiling, and bounded off the walls toward the bridge. In her local processing core, she labeled Loki as a cheater. She would update her main cores when they got back online.

BD reached the bridge to see Loki settling in one of the seats to the left of the Captain's chair. She did not know why, but she sneered at the dog in the armored EVA suit. BD sat in the Captain's chair to keep her android body from bouncing around in the upcoming fight. She started the ship's reactors, and the bridge sprang to life. Sensor data started to flood into her processors. There was an anomaly that none of her simulations accounted for. There were another five warships' behind her, all running full spectrum sweeps with static shields raised. She could not win this fight; it was time to run.

A new data point caused her processors to reallocate some of the power to new information. Havoc was in the Captain's second ship. She quickly ran more simulations with this new information. She had a plan. BD priority comm'ed Havoc's suit and the second ship to get an answer as soon as possible. After only a second, a fully nude, covered in blood, and missing his left mechanical hand, Havoc appeared on the ship's monitor. BD quickly checked the Captain's status and was relieved to see that she was alive and receiving treatment in the med pod. In less than a second, she refocused and spoke. "Havoc, we are in trouble. I need you to make a full burn to these coordinates. We will rendezvous and enter a warp bubble together." Before she could continue, EM missiles burst all around her, and the comm dropped. Humor danced through her cores when she played back the message with Havoc. In other circumstances, she decided that she would attempt some type of comment that would be considered humorous about his nudity. For now, she set the fabricator to make him a new lower arm.

Chapter 36

Havoc sat in the pilot seat and input the coordinates BD had given him. He was going to have to sling shot around the asteroid to be able to meet up with her. On his map appeared several warnings noting the locations of the mines BD knew about. "Well, there's no time like the present." Havoc maneuvered the ship out of the air lock. The whole time, his point defense cannons were kept busy by the increasing number of incoming missiles being fired from the hangar. As he cleared the hangar, the space surrounding Hel was not any friendlier. BD had tagged seven enemy ships. The space was so full of electromagnetic radiation that his sensors could not give a clear picture of the battle. He was not even sure that the Beautiful Death was still intact.

As soon as he was able, Havoc put the ship on a parallel course with the asteroid's surface and engaged the main engines. His speed built up, and he carefully plotted a course that would avoid the mines BD had indicated. In under a minute, he was able to achieve the velocity he needed to get to the opposite side of the asteroid and escape all of the ships that were currently engaged with the Beautiful Death.

As he crossed the horizon, putting the asteroid between him and the battle, his sensors cleared up, and he was able to see the rendezvous point. It was just outside the gravity well of Hel. On the rear camera, the asteroid became brightly silhouetted from multiple explosions on the far side. This time, Havoc prayed to every deity that he could think of for Loki and BD to make it out of that mess safely. Other than the prayer, there was nothing he could do but max the inertia dampeners out and get to the meet up point as soon as possible.

BD checked to ensure Loki was strapped in properly. She did not want him being thrown from the chair during their upcoming maneuvers. Once she was satisfied, she returned to the Captain's chair and strapped herself in. According to her simulations, this was going to get rough. With a powerful coded EM pulse, she activated the ordnance between her and Gus' ships. Gus' two ships fell into the trap she had modeled after their own earlier ambush. The space surrounding the ships lit up in flames and radiation. The explosion was so bright and large that she had to adjust the sensitivity on her cameras to avoid damaging her equipment. Once the explosion ran out of fuel to burn, the space cleared, and she saw, unlike their ambush, that her trap was far more successful. One of the ship's was completely destroyed, and the other was venting atmo and fire. She was not sure what ship Gus was on, but she knew that he would not be a factor for the near future.

Her simulations updated; two ships down and five ships with unknown intentions left. BD tried to comm the new warships, but the EM interference was still preventing a solid comm link. She decided that she would treat them as hostile and try to avoid them until she was forced to engage. BD turned the ship in the general direction of the rendezvous by using the maneuvering thrusters. With the acceleration she was about to engage in, it was far more efficient to be pointed in the right direction from the beginning.

The Beautiful Death was positioned to make a close pass over Hel before meeting Havoc just past the gravity well. As she started a slow acceleration, the warships began to move on an intercept course. BD turned to Loki. “That is exactly what the simulation predicted would happen. You had better get comfortable, buddy. We are about to really pull some G’s.”

“Loki ready”

BD monitored Loki’s health stats as she pushed the ship harder than any human could endure. After a few minutes, his heart rate increased, blood pressure skyrocketed, and Loki’s head began to droop. BD backed the acceleration off to a point that Loki was fully awake, but uncomfortable. Their acceleration curve was still more than a normal human could take, by far. When it was clear that the warships would not be able to catch them, a comm request came through on the Beautiful Death’s comm board.

“Beautiful Death, you are ordered by the Federation of United Worlds Warship Wasp to cut your engines and prepare to be boarded. You are wanted for smuggling illegal weapons and harboring the leader of the FWMW.” In BD’s code, there was a rule that triggered. When ordered by Federation of United Worlds officers, she was to comply as long as it was safe to do so. Luckily, Trish had blocked that fragment of her code and had given BD the option to do what she thought best.

In the current situation, BD decided to ignore the comm and power up her static shields and point defense weapons.

She kept her MAC's off line, as well as all other offensive weapons. Her programing was still very restrictive when it came to attacking government ships. She continued to out pace the warships.

"Beautiful Death, you are ordered by the Federation of United Worlds warship Wasp to cut your engines and prepare to be boarded. You are wanted for smuggling illegal weapons and harboring the leader of the FWMW. If you fail to comply, we will be forced to destroy your ship."

Humor is a strange thing, BD thought, *there is no way they can catch me. Let alone destroy this ship. That has to be a joke*. She filed it away for further investigation later. She kept accelerating until she was able to see the Captain's second ship on her sensors. She did the calculations, and at her current rate, she would catch Havoc well before they reached the rendezvous. Additionally, their delta v would prevent them from linking the ships together. "Havoc, are you strapped in?"

A surprised Havoc answered, "Yes, why?"

"I need to take control of your ship to match our speeds."

"Okay, but why do I need to be strapped in?" Havoc felt as if he had been suddenly shot out of a cannon. The inertia dampeners were over loaded, and his vision started to tunnel. He tried to reach for the control panel but could only move his left arm. and without a hand, that proved to be useless. As the tunnel closed in, the last thing he saw was a handless arm miss the control panel. BD was able to monitor the Captain and Havoc's vitals and pushed them both to the edge of their limits. BD knew that Havoc was going to be very sore when he woke up. The Captain, however, would be spared because the med pod had its own set of inertia dampeners to help offset any violent ship movements.

Chapter 37

A soft thump woke Havoc, his whole body ached as he tried to remember what was going on. After fumbling with the auto restraints, enough blood returned to his brain, and he pressed the release icon for the restraints. While the straps of the restraint system came off, they pressed into Havoc's bruised skin. The pain made him flinch, and he looked down at his chest. For an instant, he panicked. He was covered in blood and completely naked, the only thing that covered him was the emergency restraint system.

As he stood, every muscle in his body protested, he nearly collapsed back into the chair. The only thing holding him up was the mechanical arm that was jammed into the armrest. He took a deep, pained filled breath and was positive that he had a cracked a rib or two. "Great, more nanite injections, perfect." Havoc stumbled down the main corridor toward the med pod. Every starship kept an emergency supply of medical nanites onboard. As he walked, he started to remember what happened just before he passed out. He sped up as much as possible to check on Trish. The first thing he saw when reaching the med pod was his own reflection, both of his eyes were blood red, and his face was a mass of bruises.

Havoc shook off the terrifying reflection and looked at the status panel of the med pod. He took a pained breath and relaxed just a bit, all of Trish's injuries were being treated, and she would be out of the pod in 10 hours. Another thump interrupted Havoc's relief, he tried to pull up his HUD to perform a system check. However, no matter how hard he tried, his HUD would not respond. With his one good hand, he reached back and felt for his neural bridge. The component was not seated flush with his neck. He pressed the bridge back into its fully seated position, his HUD sprang to life. "What was that crazy robot thinking? Was she trying to kill me?"

"Havoc, I am not a robot, and I was most certainly was not trying to kill you. As a matter of fact, I had to slow the Captain's ship significantly to ensure that you did not perish before we met up," The voice in his head stated.

"Thanks for your concern. How long was I out?"

"12 hours and 17 minutes."

"Are you crazy?! You could have given me an aneurysm."

"You are correct. But given the circumstances, the alternatives being death, or capture then death. I am certain that I made the correct decision."

"What are you talking about? There was no way those guys could have caught up with us."

"That is true, but the ones behind us were not the problem. We were almost intercepted by the rest of the special patrol group lying in wait closer to the warp point. If we were not already going at the speeds we were, they would have been able to complete the ambush. They had 5 other warships of various sizes between us and our warp point. I had to take evasive maneuvers while accelerating, which is where you sustained most of your damage. I would suggest that you get into the med pod aboard the Beautiful Death. You did sustain some internal injuries."

"That's probably not a bad idea, but what's happening now?"

"For now, we are safe. I have linked the ships together, and we are about to enter the warp bubble." Havoc felt a second of vertigo, but did not know if it was them entering the warp bubble or symptom from his injuries. Havoc heard another thump and then felt his ears pop.

"Loki down." Havoc looked around before spotting Loki's head popping through the emergency docking hatch in the cargo area's floor. Loki climbed the rest of the way into the captured ship, as he approached Havoc. "Loki, Havoc hurt."

"I'm a little hurt, but I'll be ok. I just need to get to the med pod for a bit. How are you doing, bud?"

"Loki good. Loki help Havoc."

"Thanks buddy, but I think I can make it on my own."

"Loki, help up down."

"Up down? I don't understand, Loki."

"He is trying to tell you that when you come through the soft dock connecting our ships, the gravity will flip once you hit the half way point. I turned the gravity off on the Beautiful Death so the transition would be easier for you to manage in your current state. All you have to do is take a giant step and fall feet first through the soft dock. Once you hit the half way point, gravity will disappear, and you will softly land on the ceiling of the Beautiful Death. Questions?"

"Ya, I got one. Isn't there an easier way to do this?"

"No. Now watch Loki." Without hesitation, Loki walked the short way back to the hatch and dove head first through the soft dock tunnel. Havoc walked over and looked down. To his surprise, Loki was still slowly falling toward the ceiling of the Beautiful Death. As he contacted with his front paws, he was able to lower his hind legs to the ceiling before springing off and out of sight. "Loki fun."

"Show off." Havoc took the biggest breath he could before it hurt and stepped into the tunnel connecting the two ships. At first, everything was going well until he looked down. It only got worse when he hit the halfway point, and gravity disappeared. Havoc became disorientated and began to flail around. He reached for a ladder rung and missed completely, causing him to start tumbling. During his rotations, he was able to see that he was getting closer and closer to the ceiling of the Beautiful Death. Right before he hit, he felt the gravity turn on and began to pull him toward the floor of the Beautiful Death. As his momentum came to a halt, the gravity disappeared again.

Havoc was stranded in the middle of the cargo bay, slowly tumbling with no hand holds within reach. "A little help." BD smiled and kept recording the images. Havoc stuck floating in zero gravity, completely nude and covered in blood, had to be funny. She made a note to share it with the Captain when she got out of the med pod to see if she agreed.

Chapter 38

The two ships had been in their warp bubble for two full days before Trish woke up. Havoc, Loki, and BD all waited while the med pod finished counting down. As it reached zero, Trish began to wake up. BD made Havoc turn around while she draped a gown over Trish's form when she opened the med pod. After a few seconds, Trish opened her eyes and looked around. "Please tell me you three didn't get my ship blown up. I'm gonna be real pissed if I can't go back to my own room." She continued to look around the ship. "Captain, your ship is fully intact. Additionally, I have completely repaired the interior of the Beautiful Death from the shuttle collision."

Trish laughed, "I kinda forgot that happened. Alright, that's not important. What all have I missed? We seem to still be alive, so I guess we got away."

Havoc spoke up, "Right now, we are traveling to a fringe system. We're about three days from getting there. We also found out that the Federation of United Worlds is backing Ski for some reason."

"Wait, how do you know the government is involved?"

Havoc smiled. "Well, while you were sleeping"

"Watch it," Trish playfully warned.

"As I was saying, ten Federation warships showed up while we were escaping. BD almost killed me, and Loki may be on her side. That should be about it."

Trish sat up and reached out and took the coveralls BD was offering her. She flinched when she stretched the new skin and organs. Trish inhaled deeply and grabbed the clothes. She dropped her gown and looked at her side. A new fist sized pink scar decorated the front and back of the right side of her body. As she was poking at it, she realized that the room had gone silent. "What?"

"You're kinda naked," Havoc said, finally looking away.

"You're going to need to grow up if we are going to keep working together. Every single time we get into the mech suits, we will be naked. Get used to it."

Havoc forced himself to look at her and to keep his eyes on hers, "Okay, sorry. How are you feeling?"

"Sore. Itchy." She slipped on her coveralls and started to stretch out, "And hungry. Let's go back to my ship and get some food. I'm gonna need to see all of the data for the last couple of days. I have a feeling we're missing something. Not sure what it is, but something doesn't seem right about that task force. I know the rotations for all the task forces in this sector, and there is nothing scheduled to be anywhere near here for a couple of months. I will reach out to some old military contacts and see what I can dig up."

She stepped into her boots and headed to the open hatch in the floor. She stopped, "Hey Havoc, where is my neural bridge?"

"I don't know, where was the last place you had it?"

"It was in my suit."

"Well then, it's probably still there."

"You didn't clean out the suits?" Trish growled.

"Trish, I don't think you realize this, but I just got out of the med pod on your ship about an hour or so ago."

"What? What happened to you? Last I saw, you were almost to the ship with your suit intact."

"Your robot tried to kill me."

BD mimicked, clearing her throat, "I am not a robot, we have covered this. I believe you are doing that on purpose."

"Just like how you tried to kill me on purpose," Havoc sniped.

"That is not what happened. And I told you," BD tried to pout.

Trish interrupted, "ANYWAY, where are the suits? We have a lot of work to do."

"They are secured to the bulkhead in the Beautiful Death." BD walked over to the hatch and disappeared into the soft lock tunnel.

"Right after we eat, it's cleaning time. And it's gonna suck. I bet the blood has really started to smell." She looked at Havoc and stepped through the hatch. As she descended, she looked up at him and shouted, "You are going to help clean up the mess, it's your fault!"

"My fault? How is this my fault? I dragged you out of that suit and stuck you in the med pod, all while fighting a literal army." She was already pushing off the ceiling of the Beautiful Death's cargo bay. Loki nudged Havoc out of the way and jumped head first through the soft lock. "Loki fun."

"Yap, the dog is on the robot's side, and they are trying to kill me," Havoc announced over the comm as he dropped through the soft lock. By the time Havoc bounced off the ceiling and landed back on the floor of the Beautiful Death, Trish had already retrieved her neural link from her bloodied suit.

"Havoc, how do you still have your left arm? From the looks of your suit, you should not have all of your prosthetic."

"BD made it for me while I was in the med pod. Come on, let's head to the kitchen and get some food. We can fill in the gaps once you review the reports on your HUD."

"Okay, but you are still helping me clean out my suit while we come up with a plan."

"Fine, but would it kill you to say thanks?"

"It won't kill me, but I might kill you if you keep mentioning it, sweetie."

Chapter 39

Trish meticulously took apart her suit. She inspected, cleaned, and repaired each component before setting it aside. Trish was going to make good on her promise to make Havoc clean her suit until she saw the state of his suit. She was furious that he had not cleaned and repaired his suit before he went into the med pod. Trish's tirade lasted until he was half way through the disassembly of his suit. When she finally stopped describing how he would better serve humanity as a target for the blind, rather than a mech operator, she started working her plans for the next system.

Havoc filled her in on some of the details that were not in the reports as the pair planned and cleaned. Trish's suit was in worse shape than Havoc's, but she was done before he was halfway complete with his. When Trish finished putting her suit in its cradle, she walked over to Havoc and started pulling parts off his mech suit, expertly wiping them down and inspecting them.

"So, Havoc, do you think this will work?"

"Trish, I have no idea. If you're right about the task force being out of position, what does that mean?"

"It could mean a lot of things. I'm just not sure. When we drop back into real space, I'll reach out and get more information. For now, we need to keep our heads down."

"Captain," BD interjected as she brought parts off the fabricator, "I am not sure that will be possible. I ran a scan on your second ship and I detected an unknown subspace signature. It looks like it is broadcasting our location on a very low band channel."

"And what exactly does that mean?" Havoc asked while struggling to replace his left shoulder pauldron. Trish grabbed it out of his hands and pushed him out of the way.

"It means that someone is tracking us. And more than likely, they have been for a while." She pushed on the joint covering until the pauldron snapped into place. "BD, why didn't we pick up on this before?" She moved to the next piece of Havoc's mech suit laying on the deck, while she stared through the wall.

"It broadcasts randomly and only used a small amount of power. I would not have detected it if I had not been doing maintenance on the second ship's power systems," BD answered as she placed the new parts of the mech suits on the deck next to Havoc and Trish.

"That would explain how so many people knew we were at Hel. But it doesn't explain the government involvement." Trish picked up the newly fabricated lower left arm piece to Havoc's suit and absentmindedly inspected it while she tried to make sense of the events of the last couple of weeks.

"So, if the new ship is broadcasting our location, why don't we just get rid of it?" Havoc said while struggling to snap on a section of his back armor.

"Because, now that we know we are being tracked, we can use that to set up the next engagement to our advantage," Trish mused while working on a connection on the front of Havoc's suit.

"Trish, why does there have to be another engagement? Why can't we cut the second ship loose and make a run for it?"

"Havoc, you are thinking like James. We both know what happened to him. There will be no running from this while we still have the XS-12's. People have been killed just for having known about them. No, we have to figure out who is really pulling the strings, or this will never end."

Havoc stopped what he was doing. That was the first time she had ever compared him to his former self; and for some reason, it hurt him. He took a breath to calm himself, "Trish, remember, I'm still very new to this. It's only been a couple of months since this whole thing began. I think I'm doing pretty well considering I haven't once completed a single training cycle for space or ground combat," He started grinding the words out by the time he was finished talking.

Trish stopped what she was doing and looked around Havoc's mech suit so she could see him. Now it was her turn to take a breath, "Havoc, I'm sorry. You do so well at certain things, and I sometimes forget that this has not been your whole life. It seems like the last month or so has been wearing on me more than I thought."

Havoc was stunned. He had never heard her give a real apology. Not only that, but this was the first time she acted like a normal person around him. "It's ok, I guess we are both on edge."

Havoc started to relax and he felt calmer than he had in a while. He looked over at Trish and noticed that she seemed more relaxed than she had been just a few seconds ago.

"Captain, I am detecting a foreign substance in the air. The levels are spiking, but I do not believe it is harmful."

Havoc looked over at Loki. His big golden eyes were darting between Trish and Havoc as he sat next to the pair.

"Loki, can you take it down a notch? We are fine. We just needed to get some stuff off our chest."

"Loki, help."

Trish looked between the pair, "What are you talking about? Help what?"

Havoc laughed. He never remembered telling Trish about Loki's super power. "Loki can produce a counter pheromone to stress. And I would bet 100 credits that he just pumped the hangar full of them."

"Loki, help."

"I want to be mad, but I have not felt this relaxed since I met you. No offense," Trish resumed cleaning and replacing parts on Havoc's suit.

"None taken, I forgot how nice it is when he does this." Havoc finished snapping in his back armor into place. The pair worked in silence for a while and enjoyed the temporary peace.

"BD."

"Yes, Captain?"

"How hard would it be to find the device that is broadcasting our location?"

"It would take some time to trace the location of the power drain, but I am confident that I will find it within the next three hours if I start now."

"Get started on that. I have an idea, but I need to know if you can remove the device without its owners knowing that it has been tampered with."

"Captain, am I to understand that you do not want me to disable the device?"

"Correct, I have plans for that little guy."

"Do you mind sharing?" Havoc asked.

"Well, everything depends on if we can move that transponder, or whatever it is, without alerting anyone."

"Ya, I got that. But what comes next?"

"If we can move it, then we will have two ships in the fight. If not, well, then we will have to use Gus's Folly as a fire ship."

"Um, I have a couple of questions. First, what is Gus's Folly? And the second, what is a fire ship?"

"Easy," Trish quipped, "Gus's Folly is the name of my second ship."

Havoc interrupted her, "Why is the second ship yours? And what is your deal with naming ships?"

Trish narrowed her eyes and stared at Havoc as she continued, "I get to name MY second ship because I was the one who went to the rogue planet and stormed the ship by myself."

Loki's ears perked up, "Loki, Help."

Havoc chimed in, "That's right. Loki was there with you and helped you. And since he is my dog, I should at least be part owner."

Trish exhaled loudly, "I will concede that Loki was there and helped. So I will make him part owner of the ship, not you. Now, Loki, is it okay if we call our new ship 'Gus's Folly'?"

"Loki, yes."

"Traitor," Havoc muttered.

"Anyway, now that that is settled, a fire ship is a large wooden vessel that was set on fire to be used against enemy ships during a ramming attack back when wooden ships fought battles. Or, in our case, a fire ship is a spaceship with an XS-12 on a remote switch."

"Oh, that seems pretty straight forward."

"Simple plans are often the best. Especially when we don't really have a crew and will almost certainly be out gunned at least five to one."

"Okay. So, what is the rest of the plan? There has to be more than that."

"In three hours, we can start working on a more refined plan. For now, let's hit the VR trainer. You have a lot to catch up on, and this is going to be fun."

Now Havoc was scared.

Chapter 40

After several hours and multiple deaths, Trish finally let Havoc out of the VR trainer. It had been three weeks since they were able to remove the locator beacon and power it up without its signal being interrupted. Every few days, they would drop out of null space and measure the emergence of the fleet chasing them. On average, the fleet would emerge within three hours after them; and within 2 light seconds from the location the Beautiful Death and Gus's Folly emerged from. With each jump, Trish refined her plan until she had set all her pieces in place for the confrontation.

Havoc and Loki did not get to rest. While Trish was planning and scheming, they were training. Whether it was physical training in the real world, vac training in the opened cargo bay, or combat sims in the Beautiful Death's VR trainer. Every minute of their waking day was accounted for and used to make the pair a lethal combination. By the end of the third week, Havoc could operate both ships, clear starships, and maneuver around the void comfortably by himself. If he was teamed up with Loki, they could almost hold their own against Trish and BD.

Havoc removed his goggles and unhooked from the VR's neural link. His HUD disappeared and he quickly put in his bridge. He felt part of his senses cut off when he did not have his HUD up. Havoc sat in the middle of the VR room and watched the sweat pool as it dripped off his face. Trish walked up to Loki, removed his helmet, and rubbed his ears.

"Well, Havoc, I think we are about as ready as we can get. I have one more pick up, then we are going to set up the ambush."

"What do you mean 'one more pick up'? When was the first pick up? And what are we picking up?"

"Havoc, you went out into the void and grabbed most of the gear yourself. Please tell me you're kidding."

"I have no idea what you are talking about. The only thing I did in the void was move debris from outside the ship as part of my training."

"And where did you put all of it?"

"You told me to bring it into the hangar, but that was all junk."

"Do you see the junk now, Havoc?"

"Well, no. I just figured you vented it into space when I was done with the training."

"Okay, I guess I will spell this out for you. Back in my special forces days, we planted supplies all across the galaxy in case we need a covert resupply. We masked its signature and disguised it to look like worthless junk. That is what you have been pulling in. I stashed all of the gear in the troop berthing's armory, or in the rack spaces. That's why they're not here out in the hangar bay anymore."

Havoc sat there, completing the mental puzzle he had just been handed. "Well, where was I when you moved all of this?"

"That's what you care about? Not what was in the boxes? Nothing about special forces? But when did I put it away? You've got to be kidding me. Listen. Tomorrow, we drop out of our warp bubble. We have been making a pretty predictable path to a known pirate base. I'm betting that the people chasing us are going to set up an ambush there."

"And the people chasing us are part of the federation. Doesn't that complicate things?"

"Look, Havoc, I don't think they are. Every resource that I have says that the fleets are where they are supposed to be, even the black ops ones. I think they are just masquerading as feds."

"But that would mean Ski has a whole army at his disposal. That, or Gus, is way more powerful than we thought."

"It's the first one. Gus is not a real player in this. He is just a greedy thief. I am betting after Ski ransomed that planet, he bought an armada of ships and filled them to the brim with mercs and ex-military."

"This won't be easy. Will it, Trish?"

"No. This is going to be very hard, and I am not sure we can pull this off. But tomorrow is our best chance at winning this thing. If we wait much longer, I think that Ski will be able to gather more forces. Then we won't have a chance at all." Trish stared into cargo bay, looking directly at the XS-12's stacked neatly in the corner. "Let's get some dinner and head to bed. Tomorrow may be a very long day."

It was three hours after they went to bed when Havoc's door slid open. Trish stepped through wearing a pair of bright pink pajamas decorated little cartoon depictions of explosions and mayhem randomly placed across the entirety of the long shirt and pants. Havoc was not quite asleep and opened his eyes when the light from corridor lit up his room. Loki was curled up at Havoc's head between him and the wall. He had put his chin on Havoc's forehead and stared at Trish. Trish cleared her throat, "Don't get any ideas. I can't sleep. I thought Loki could pump the room full of his relax gas and help. Scoot over and make some room." Trish plopped into the small bed, forcing Havoc and Loki to adjust to give her room. "What the hell, Trish? You can't just barge in here like that."

"Why not? It's my ship," She said, pushing her way to more space.

"Ya, but this is my room. I could have been doing something personal in here."

"With Loki in here, that's not personal. That's gross."

"Okay, Trish, that's a fair point. But still."

"Relax, I checked your vitals before I came over here. Now, Loki, crank up the relaxing vibe. I want to get some shut eye." The three of them were fast asleep before the conversation could go any further.

Chapter 41

Havoc and Loki rocketed across the asteroid field at the best speed that Gus's Folly could make. They just dropped the core buster on the largest asteroid they could find, and Trish was on the Beautiful Death with the tablet with the abort codes. The pair had placed the transponder and its power supply next to the weapon. Now, Loki sat in his EVA suit clipped into the copilot seat, watching as the occasional asteroid flashed by. A flash lit up the cockpit, quickly followed by so many additional flashes that Havoc had to rely on the computers in Gus's Folly to get an accurate count. Ski's fleet had just dropped out of null space, and it had nearly doubled in size to 17 ships. Trish was right, it was now or never.

The Ski's fleet showed up two and a half hours early, and well out of their normal emergence point. The fleet had separated into three distinct formations and had enough separation that a single XS-12 in open space would not be able to destroy more than one formation. Havoc almost panicked. This was not what they planned for. A tight beam laser comm request appeared on his HUD. Havoc accepted the comm request and his HUD updated with the combat net from the Beautiful Death. The squadrons were tagged; Alpha had 7 ships, Bravo had 5 ships, and Charlie had 5 ships, all arranged in a V formation.

Havoc jumped when BD's voice sounded in his head. It sounded to him like it had lost its innocence. In the last couple of weeks, she had developed a hard edge to her voice. It unsettled him. "Havoc, we have a change of plans. We need you to get to the following coordinates at best speed, then shut down Gus's Folly completely and wait for further instructions." The connection cut out before Havoc could ask any questions.

An unencrypted, open comm broadcast interrupted the silence of space. It was Ski, and he was holding something that looked oddly familiar to Havoc. "James Minder, I know you are out there. I can see you darting around in that asteroid field. Why don't you come out? Stop messing around and talk to me. You don't have to die for real. I can make you an extremely rich man, and you can stop playing bounty hunter."

Havoc shifted uncomfortably. How could he possibly know that he was James Minder? His team did everything perfectly. Havoc swallowed hard and opened up his own comm, voice only. "This is Jack Havoc, what do you want?"

"James, is that you? You do sound different. Come on, turn on your vid. I want to see you."

Havoc took another breath and switched on the ship's camera, "I'm not sure who are, but what do you want?"

"Who programed your gene mod? They did a fantastic job. I may need them in a little bit."

"Look, guy, I am not who you think I am-" Ski cut Havoc off.

"How is your left arm?" Ski asked, Havoc froze in place. Ski waved with the black and orange object he was holding. After a second, Ski used the object to scratch his face. Reality slapped Havoc so hard, he forgot to breathe. The object Ski was using to scratch his face was James Minder's arm. The one that was torn off when he was making his escape from his old recon ship. Ski continued, "It was crazy. I had to verify whatever Sergeant Keller was saying. So, I sent a team to look at the body and inspect what was left of the cargo. I had been completely suckered. I was about to let the whole situation go. Then, one of my sensor techs happened to catch a blip floating on her sensors. Me being the curious man that I am, I sent a shuttle to investigate. And can you believe that it was an arm just floating through space? Now, that got my curiosity really going, so I had it tested for the DNA profile. And it came back as belonging to the late James Minder. Odd thing, though, James Minder was missing his head, but had both of his arms. Weird, right? So, back to my original question, how is your left arm?" Ski's ice blue eyes stared at him through the monitor.

Havoc swallowed hard, "What do you want, Ski? I am getting tired of this game."

"Mr. Minder, what I want is all of the XS-12's you have left. Which, by my count, is four. I don't want to hear that nonsense about you having more. I know you don't have anymore. So please don't play that game again."

"And what do I get out of this deal?"

Ski chuckled, "You get to live. And if you quit stalling, I will give you the full price that you paid me. Even though you have already been paid once. But try anything cute, and I will kill you, your dog, and everyone you have known." Explosions ripped across the squadron designated Bravo. After the radiation cleared, Havoc saw only two of the five starships in Bravo were still intact.

“Havoc, we have a change of plans. We need you to get to the following coordinates at the best speed, then shut down the Gus’s Folly completely and wait for further instructions.” BD cut the laser comm before her location could be traced by the newly arrived squadrons. Beautiful Death slid through the black, completely powered down on a preset path that kept the asteroid field between her and the enemy squadrons. The passive sensors gathering the data she needed as she waited to see Captain Keller’s signal.

BD threw stealth mines out the open cargo of the Beautiful Death in a random pattern along the outer rim of the asteroid belt as the ship costed along. She continued to throw them until she ran out of the munition. Now, she had to wait for the Captain to give her the signal to move on to the next phase. She moved to the bridge. All of the ship’s feeds went to her processors. She had a complete, moving tactical map of the solar system as she sat motionless in complete darkness.

BD sat there as Havoc and Ski talked over the open comm’s. When Ski announced that it was James Minder’s arm he was holding, BD’s processors started to heat up as she tried to calculate the probabilities that the arm was really James Minder’s. While she was calculating that, she was also creating a predictive model on Captain Keller’s drift through open space. She did not have time to finish all the math before the explosions ripped through Bravo squadron. When the radiation cleared, three of the ships were completely destroyed and the other two had minor damage.

With the signal given, BD accelerated the Beautiful Death to the point where the ship was almost coming apart. Any living thing, to include Loki, would have been crushed to death by the g force within the first seconds as she burned hard to get above the elliptical plane of the solar system and raced toward Bravo squadron. She powered up her static shields, charged all of her weapons, and pushed all her sensors to max. She would be lit up like a torch on the enemy sensors. Nothing she could do about that now.

Trish had just finished positioning the dormant missiles in the spot where BD had left her, when the flash of q-drives startled her. She was not surprised that they were early. She was surprised that they appeared all around her and so close that she could see at least four ships with no assistance from her mech suit. With assistance, she could make out three formations totaling 17 ships. "Great, this changes things. But, the good news is that I can throw these missiles and hit a couple of the ships."

Trish's suit picked up an open comm coming from of the ships and decided to listen in. "James Minder, I know you are out there. I can see you darting around in that asteroid field. Why don't you come out? Stop messing around and talk to me. You don't have to die for real. I can make you an extremely rich man, and you can stop playing bounty hunter."

Trish listened as the conversation went on. She used her cold jets in short bursts to maneuver the missiles into the right position. She was close to reorienting the last missile when she heard Ski talk about finding James' arm and running the DNA profile. "Well damn, that explains a lot. No time like the present, I was starting to get bored anyway," Trish said to herself.

She powered up her suit just enough to give the launch command to the missiles. She smiled as the munitions rocketed toward their targets. She programmed the missiles to hit evenly across the squadron closest to her, except the ship that was closest. She only programed a single missile to hit that ship. She counted to three in her head, then powered up her suit and hit a full burn on her maneuvering thrusters to follow the missile in. Trish's smile grew as she watched the missiles hit the hulls of the starships and blossom into the beautiful yellow and reds of destruction. She had to let the team know that she was no longer in an open space. Trish opened an encrypted all channel broadcast. "Change of plans. I am playing with B4. BD, you have the toys. Havoc, stay the course." She made herself focus on the task at hand as she sailed through the newly created hole in the starship she designated B4.

Chapter 42

Havoc did not even make it to his designated shut down location before the explosions tore across the Bravo squadron. The large set of explosions was the signal, according to the plan. Now, he was supposed to burn hard toward the largest group of starships. Trish's voice interrupted his thoughts, "Change of plans. I am playing with B4. BD, you have the toys. Havoc, stay the course." The comm died. Havoc's sensors told him that the two remaining ships from Bravo were making a run for Charlie. He had to guess that Trish was somehow in one of them from the cryptic message she just sent. "So, the attack drones that she was supposed to ride on to get to the pickup point are now under BD's control," He muttered, "That means I should point the ship at Alpha, and I will still be in the right spot for pick up."

Havoc decided that it was time to get out of the cockpit and get into his Mech suit that was back in the cargo hold. As he ran through Gus's Folly, he was a little sad that the ship and all was going to be destroyed. They did not even have the time to remove the valuables out. Once he got to the cargo bay, he spotted XS-12 number 8 strapped to the wall. It still gave him chills just looking at it. He continued to his mech suit with Loki in tow. Once he got there, he stripped down and placed his jumpsuit in the small leg compartment before climbing in. Loki watched as he climbed into his mech suit. Havoc squealed when his suit connected to him. "I will never get used to that," He said before the suit closed over him, "Never." This time, it was over the local comm net. "Okay, Loki. This is the rough part."

Havoc walked over to the cargo door and remotely emptied the ship to the vacuum of space. Loki trotted over and jumped onto Havoc's chest. Havoc used the suit's tether to secure Loki's armored suit to his chest. After the cargo bay was depressurized, Havoc opened the cargo doors and looked out into the vastness of open space. He sent the Beautiful Death a tight beam comm through Gus's Folly's systems, "Walking the dog." He cut the comm. Havoc jumped out as far as he could with Loki strapped across his chest. His guts twisted as he left the artificial gravity and tumbled through space. Luckily for him, his suit was able to arrest his tumble within a couple of seconds. Havoc was able to orient himself to see Gus's Folly's engines flash white hot as BD took over the ship's controls remotely.

All he could do now is wait. "I still don't see why BD could not have remotely piloted that ship from the beginning." He tried to mimic Trish's girlish voice, "Because, Havoc, there are too many variables and unknowns for that ship to be unmanned." Havoc thought about for a few minutes before asking Loki, "Do you think we're bait?" He could hear the mirth in Loki's voice when he replied, "Loki, yes."

BD was waiting for the ten assault drones to come online to signify that Captain Keller had begun her portion of the attack. As the Beautiful Death reached the apex of its fight path, BD received an encrypted all channel comm. "Change of plans. I am playing with B4. BD, you have the toys. Havoc, stay the course." BD 's processors heated up as she went through all of the possibilities of what the message could mean. She almost instantly ruled out the possibility of the comm coming from an enemy ship. It was her encryption codes that came across the net and there was only a statistically insignificant probability that her codes were broken on their first use.

The message had to be treated as authentic. As for the meaning, her algorithms predicted that Captain Keller had changed her course of action, and she was now on the ship designated Bravo 4. BD had estimated 97.7% chance correctly designating what ship the captain was on after analyzing all available data. Once that variable was eliminated, she moved on to the next part of the message. She was in control of the toys, the attack drones. She pinged them, and they responded within the acceptable timeframe. She programmed nine of the ten attack drones to fire at the ships she planned on damaging in the Alpha squadron. The attack drones would arrive 20 seconds after she made her strafing run.

The last part of the message involved Havoc staying the course, which meant that Havoc would be abandoning Gus's Folly after positioning it to become a fire ship for Alpha squadron. By her calculations, that call would come within the next five minutes. BD ran multiple simulations on how to achieve the best outcome given the current conditions. Now, she had to alter the plan. BD sent the last drone to Havoc's predicted location. Gus's Folly sent a tight beam burst comm with the same encryption codes that Captain Keller just used. "Walking the dog." Havoc's voice notified her that he was floating in space, waiting for pick up, and Gus's Folly was hers to control.

BD updated the attack drone that she was going to pick up Captain Keller with. She programed in Havoc's vector and message to be delivered, then the drone would under be his control. After that was complete, she committed all her processing power, her attack run, on Alpha squadron. She oriented the Beautiful Death on the largest ship and fired a salvo with MAC's at its predicted location, the time of flight for rounds to impact was two minutes. BD overrode the ship's safetys and red lined the reactor cores. She needed all velocity she could get in order to prevent Alpha squadron's missiles from tracking her down after she passed through their squadron. She devoted a tiny sliver of her processing power to watch the ship designated Bravo 4 as it joined Charlie. BD updated her simulations as Charlie squadron turned toward a converging heading on Beautiful Death's projected path. Their new heading would take them away from the buried XS-12 and have them intersect Havoc's vector.

Trish rolled as she hit the deck of Bravo 4, coming up in a shooting stance. She searched for targets in the passageway. The artificial gravity was still working, but this section of the ship was open to space. Nothing moved, so she decided to press on. She moved to the closest bulkhead and planted a breaching charge. Trish ducked into a small alcove that was used for storage and detonated the explosive. When the charge exploded, she could not hear it. But she did feel it through the ship and her suit. As soon as the initial rush of atmo was gone, she ran into the newly opened section of the ship.

This time, the passageway was not empty. There were several unprotected bodies that were still in the process of dying, and even more that had expired. None of them were wearing a uniform or vac suit, further enforcing Trish's theory that these were not government ships. Trish did not have time to examine the bodies further, a string of low velocity rounds bounced off the back of her mech suit's armor. Trish's HUD indicated the direction the rounds came from and she charged down the corridor.

Five men in lightly armored EVA suits had been behind a make shift barricade when she had breached the wall. Now, they were firing in two teams to ensure that they kept her pinned down. Trish barely thought about it before she charged head strong into the ambush. Her armor would hold up to the low velocity slugs abuse it was receiving, but she did not want to push her luck. Trish raised her S&K assault rifle and let loose a full salvo of plasma core rounds into the barricade. The rounds easily punched through it, the men hiding behind and into the bulkheads behind them.

Trish smiled when all return fire stopped. "Next time, maybe you should use the right rounds when fighting a mech suit." She ran past the bubbling puddles of blood and started to make her way to the engineering compartment. She noticed strobing red lights and just assumed if there was any atmo in there, she would have heard a wailing siren to indicate the ship had been boarded. "Well, they know I am here now. This is going to be fun." She switched her suit color from the blackness of space to the bright pink she favored when she was not trying to be stealthy. She drew her mono-blade short sword in her off hand, clipped the rifle to her chest, and pulled her plasma core pistol. Now, she was ready for the close in fight.

Chapter 43

Havoc and Loki floated in the emptiness of open space. The first couple of times Trish had him go into space without any tether, he almost had a panic attack. But she had made him do it so often in the last three weeks in VR and real life, it was almost routine now. Havoc kept himself orientated to where he thought the battle would be taking place, watching for the slightest flicker of light or a star suddenly being blocked out.

A small ball of flame erupted about where Havoc thought it would. If BD's math was right, he should be getting picked up by the Beautiful Death in five minutes. A proximity ping alerted Havoc to something approaching from his 9 o'clock high. He gave a quick burst to his maneuvering thrusters, then a counter burst when he was squared up on the object. He could not see anything, but his sensors said the object was closing the distance and slowing down. Havoc already had his rifle out when the low power comm notified him that it was one of his assault drones. That was not in the plans, he was supposed to get a ride from the Beautiful Death.

As the drone approached, he received a text only comm. The message floated in the center of his HUD, "Attach Loki and yourself to the drone, plug in the data cable for further instructions." "This is just perfect. Come on, Loki. Let's see what surprises await us next." Havoc unclipped Loki from his chest and waited for the drone to come within an arm's reach. He clipped Loki in, then clipped himself to the top of the drone next to Loki. As soon as he plugged in the data link cable, the drone began accelerating away from the explosion that Havoc saw earlier. The drone started to accelerate so fast that the tether was the only thing stopping him from being thrown from the drone's top and sent tumbling into space. Havoc looked at Loki, he was lying on his stomach with all four suit paws magnetically locked to the drone. Havoc dropped to his stomach, dug his mechanical fingers into the drone's skin, and held on.

Information started to stream over his HUD through the data cable. Just as Havoc thought he was getting a handle on the situation, the Beautiful Death appeared from behind him out of the inky blackness of space. Before Havoc could react, the ship was nothing more than a faint star in the distance, moving in an arc across his field of vision. Havoc blinked, and when he opened his eyes, he was surrounded by a seemingly endless sea of floating metal foil pieces and his HUD blanked out for a second before his suit could adjust to the new EM environment. For the first few seconds, the foil pieces were passing him at an incredible rate, but as the assault drone he was riding continued to accelerate, the pieces seemed to slow down until they looked like they were frozen in place.

The drone cut its engines, and gravity disappeared as he lay there trying to parse through all of the data BD had sent him. The space around him erupted with missiles burning hard after the Beautiful Death. Some of them so close he swore that he could have reached out and touched them. While the missiles were accelerating through the foil, they seemed to lose track of their target for a brief instant before reorienting and burning to chase their prey. Havoc franticly checked over the data to see what his next part in the new plan was. He got to the part about incoming missiles, as the only traces he could see of them was the glow of their reactor engines. Havoc sped through the rest of the data as quickly as possible. From what he understood, he had two courses of action; take the drone and hide in the asteroid belt, or use the drone to board Bravo 4 and assist Trish. "I was wrong, buddy. Now comes the rough part." "Loki ready." Havoc tried to match the vectors of Bravo 4 with his limited data. Any second now, he would be overtaken by Charlie with Bravo 4 in trail. Now, he just had to find the ship and hope it was the same one Trish was on.

BD watched as the largest ship burst into flames and split into multiple pieces. When the MAC rounds impacted, she adjusted her target and fired the MAC's again. The chances of her hitting a second target were low due to the squadron's evasive movements. The MAC rounds had a one-minute time of flight and she would be in the middle of the squadron 30 seconds later. With their delta-V, she would pass through the squadron from top to bottom while being in their firing range for 10 seconds for kinetic rounds and an additional two minutes for the smart munitions.

BD used the opening in the formation left by the ship she had just destroyed and pushed the Beautiful Death toward it. At their maximum effective range, she fired a full salvo of hyper velocity missiles across the remaining six ships. Once the missiles were on their way, she diverted power to the forward static shield. She would adjust the power to the mid shields and aft shields as she passed through the formation. Once she was in range, she opened up with her point defense turrets. Fire engulfed her as she passed through the formation. She could sense temperatures rising on the skin of the ship due to the amount of ships firing at her. She was taking limited damage and lost a port lidar sensor as she passed through, but she would be able to cope with that.

Beautiful Death's aft static shield lit up as it struggled with the amount of impacts it was absorbing. Her battle damage assessment was two ships destroyed and two with moderate damage. The speed difference between the two belligerents bought BD enough time to enact the next part of her plan. She located the assault drone that she had sent to Havoc and set a course to pass just close enough not to burn him with her engine plumes. As she closed the distance, she loaded chaff and EM missiles into her forward tubes. When BD reached the correct distance, she fired the missiles in Havoc's direction to mask him from the enemy missiles and ships. She adjusted her turn to give Havoc the best chance of boarding Bravo 4 as they tracked her on their pursuit course.

BD adjusted Gus's Folly's course to pass well above the current battle. With the unpredictability of the XS-12's, she did not want to risk having Captain Keller, Loki, and Havoc pulled into the singularity or blown apart when it exploded. She would have to save that surprise for a different engagement.

BD watched as Havoc's location was lost in a wash of sensor interference, then turned her attention to the assault drones entering the firing range on Alpha squadron. The drones focused their fire power on the already damaged ships. Three of the drones were destroyed shortly after losing their missiles, and the squadron point defense turrets managed to destroy most of the incoming warheads. Several missiles did make it into the opened sections of the damaged ships. One of the two damaged ships suffered a reactor failure and had to eject its core to prevent losing the ship. The second starship was venting atmo and flame into space and started to fall behind the squadron. The remaining six drones turned to pursue Charlie squadron.

BD watched as the missiles fired by both Charlie and Alpha squadron streaked past where she predicted Havoc to be on a pursuit course for her. Not a single one exploded anywhere close to Havoc, however, several came very close to his location. BD calculated Charlie squadron was now almost on top of Havoc. Now, it was up to him to choose fight or flight

The ship was much larger that Trish initially thought. She would have to be very careful not to get boxed in. Her suit could take the low velocity rounds for an extended amount of time, but she doubted that the enemy would use those rounds for much longer. The ship's layout felt very familiar, almost like one of the old decommissioned training ships her spec ops team used for practice boarding operations. If that was the case, there would be an emergency internal ladder running through all the decks of the ship.

Trish changed her direction and started working toward the location of where she thought the ladder would be. So far, she had only encountered light resistance, and bodies that died due to the exposure to vacuum. This was definitely not a military ship. Ski had somehow purchased a fleet and manned it with poorly trained personnel. Trish rounded the corner at an intersection of corridors and was immediately hammered with a salvo of plasma core rounds. The impact was so great that it knocked her on her back for a second. Her chest armor flashed a yellow warning in her HUD. She would not be able to take another salvo like that.

Her training kicked in, and she rolled back to the cover of the corridor she was just in. She quickly thought about how she would finish the ambush if the roles were reversed. Behind her, they would be coming from behind her. She turned and threw a breaching charge on the deck just far enough not to be damaged in the blast. As soon as it exploded, she jumped to the ceiling and magnetically and mechanically locked her suit in position. She then activated her active camouflage and then waited. While pressed against the ceiling, she looked like just another beam that ran through out the ship.

Just as she planned, the group that was trying to sneak up behind her and the group that had her pinned in place started to come around opposite corners of the corridor. Neither group looked up, instead staring at the hole in the floor that led one deck below. Both fire teams had one heavy mech suit in them. Trish knew she would have been in a tight spot if she was caught between the two fire teams. With four members in each team plus two well-armed heavy mech suits, she would have to ambush them or avoid them. With only two places for a boarding party to go, the enemy would be waiting for her at the bridge and engineering. "Ambush it is," She muttered to herself.

She patiently waited for her chance. She had not cracked their comm net yet, but by the way, they were pointing around, she figured out who was in charge and had a rough idea of their plan. After a few seconds, the fire team that had come up behind her jumped into the hole in the floor she had created. The mech suit was the first to go, followed by the other four members of the fire team. As soon as the last one was gone, she carefully removed another breeching charge and threw it onto the back of the mech suit, and immediately detonated it.

The breach charge destroyed the top back half of the mech suit and the body inside it. Before the suit hit the ground, Trish was already on the move. She kicked off the ceiling and landed in between the four surviving members of the fire team. Trish drew her mono-blade across the closest man's chest and nearly cut him in half. On her right, she fired a controlled pair right into the face plate of the woman farthest away. The first round burned through her face plate and blinded her, the second plasma core round penetrated her skull, completely destroying her brain. Both bodies hit the ground at nearly the same time, blood spilling onto the deck then boiling away in the vacuum.

The last two members of the fire team had started to raise their weapons. Trish quickly threw her mono-blade into the neck seal of the man farthest away. The blade easily passed through the flexible neck armor and severed the spinal cord before the hilt of the knife stopped the blade from going any further. Trish fired at the last man, but he was already dropping out of her bullets' path. From his crouched position, he sprang forward, putting his shoulder into Trish's stomach, and tackled her to the ground, Trish let the momentum carry them to the deck. As she fell, she wrapped her legs around his rib cage and used the mechanical power from her mech suit to squeeze until she felt his ribs shatter. She then brought down her armored elbow onto the top of the man's helmet over and over again until it was so deformed that the face plate seal failed. The man stopped fighting her and began to thrash wildly as the atmosphere was pulled from his suit. Trish rolled the still spasming body off her.

She stood up and surveyed the carnage. Trish then walked over to the man with her knife sticking out of his throat. She calmly retrieved the mono-blade from the corpse and flicked the boiling blood from the blade onto the twitching body. "On to the next." In two steps, she disappeared down the hole in the floor.

Chapter 44

Bravo 4 was getting so close. Havoc thought that he could reach out and almost touch it. He saw the damage from the missile and was just able to make out some of the details in the exposed corridor. The attack drone that he was riding alerted him that they would be out of the chaff cloud shortly. Once that happened, the drone would be an easy target for Charlie squadron's point defense guns. Havoc programmed the drone to crash into Bravo 4's exposed deck. "Brace, buddy." Havoc gritted his teeth as the g load increased. He had nearly blacked out as all movement came to a violent and abrupt stop.

Havoc unclipped himself and rolled off the drone onto the deck of the starship. He lay there on his back for a few seconds, trying to blink the stars that floated in his vision away. Suddenly, there was an impact on his chest. Before he could react, Loki's helmeted head appeared in front of him. "Loki, Okay?"

"Ya buddy, I'm okay. Just knocked my head a bit." Havoc started to roll to his stomach, and Loki jumped off. Getting off the ground in his mech suit was a carefully choreographed movement, getting a walking tank upright was not an easy feat. Havoc finally got his feet under him and took stock of the area. 'Mayhem' was the word that popped into his head as he looked at the damaged ship. The fact that his combat drone was wedged into the missile damage did not help. As he stood there taking the damage, it accrued to him that there was no damage control personnel trying to repair the ship. He filed that away for later. "Okay, Loki. Let's go find Trish."

Havoc's mech suit would barely fit through what he assumed was Trish's bulkhead breach. She had made him run through multiple VR scenarios where he was required to board a starship. Trish had also informed him that the reason she picked this mech suit for him was because it was the largest one that was rated for ship combat. Havoc stowed his auto cannon and shimmied through the jagged hole. Loki was already on the other side of the bulkhead, counting the bodies that were laid out where they died. After a quick glance, Havoc summarized that all of them had died due to the exposure to vacuum, not Trish's personal handy work.

Havoc looked around to see if he could figure out where Trish went. He almost asked Loki if he could smell her but then remembered while they had artificial gravity, they were in complete vacuum. As Havoc scanned the area, he noticed impact marks on one of the bulkheads. He walked over and then looked at the direction they must have come from. Further down the corridor, he saw several bodies laid out with large holes punched into the suits and crimson stains silhouetting them. "Loki, this way." That was Trish's handy work, he thought. Now, all he had to do was follow the trail of carnage.

BD was above a 99% certainty that Havoc had made it onto a ship in Charlie squadron. She had calculated a 72.43% chance that he would manage to get on the correct ship. With that much uncertainty in her calculation, she removed Charlie squadron ships from her list of targets until she could pinpoint the ship, or ships, that the Captain and Havoc currently occupied. BD then put Gus's Folly in a holding position far above the battle space. Once it was in position, she powered down Gus's Folly's systems, putting the ship into standby mode.

The six remaining assault drones were on an intersecting arc with Alpha squadron. Soon, BD would have to stress her frame past its published limits to orient her MAC's and EM and chaff missiles on Alpha squadron. The timing had to be perfect if she was going to time the arrival of her rounds with the assault drones' suicide run. She felt the ship twist around her as she experienced more g's than any known living being could withstand. In 20 seconds, she had completed a 270 degree turn at full burn, perfectly lining up her missile tubes. She fired a full spread of EM and chaff. Her auto loaders quickly refilled the empty missile tube with low yield nukes. Ten seconds later, BD fired her last round of missiles.

Just before she lost the sensor readings due to the chaff and EM interference, BD plotted all three of Alpha squadron's ships. She adjusted the ship's angle and fired the MAC's at the lead ship. The EM and chaff exploded and she lost the sight picture. BD continued to adjust the angle, and when she predicted she was in the correct position, she fired again. She then repeated the process one more time. It was 24.2 seconds between her first MAC shot and last. Now, the barrels had to cool, or she risked rupturing them.

If her plan worked correctly, the MAC rounds would erupt out of the interference cloud, not giving the ships a chance to react. The low yield nukes would be able to take advantage of the disrupted formation because the ships would no longer have overlapping point defense. In the last part of her plan, Beautiful Death would meet the enemy fleet in the disruption cloud and be able to pass the formation, firing her own point defense to cause more damage.

The Beautiful Death entered the interference cloud and lost all sensor data. She flew blind, firing her point defense cannons on the projected paths of the enemy ships. BD passed through the cloud without taking a single shot from the enemy. She was not able to even find any projectiles fired at her. When she cleared the cloud, she saw the carnage on Alpha squadron. Not a single ship was recognizable. They were reduced to rapidly cooling chunks of metal tumbling through space.

Her sensors came alive, warnings of over 100 incoming missiles from the seven ships of Charlie squadron. By her calculations, Charlie was almost 20,000 miles closer to her than where she had predicted. BD quickly ran her simulations, she had a 23% chance of surviving for the next 10 minutes if there was no drastic change to the battle space. She began the course of action that gave her the best odds of survival.

Trish was stuck between a hammer and an anvil. She let herself think that she was the only predator aboard the ship. Trish had been following the fire team that went through her deck breach when the ambush nearly killed her. The team must have been alerted because they lead her into a T intersection. She was close enough to watch them turn toward the aft of the ship. When Trish followed them around the corner, they were all facing her and opened fire. The impact knocked her flat on her back, and the chest plate of her armor flashed Red in her HUD. The kinetic shock knocked the wind from her lungs, it was only the years of training that saved her.

She rolled out of the kill zone, only to find another group of mercs coming from the bow section of the ship. She quickly got to her feet and started to run in the direction she had come from. The shots to the back of her armor continually made her stumble and trip as they slammed into her. After a few seconds of plasma core rounds fired from the mech suits, her entire torso armor was flashing red in her HUD. The corridor was too long, her armor would not last long enough for her to turn the corner. In a last desperate act, she pulled her last breach charge, turned, and threw it at her attackers. Trish's world slowed as she watched the charge seemingly float as it arced toward her attackers. She saw the fire team try to react by back peddling, but they were blocked by the mercs behind them. When she thought the charge was close enough to her attackers, she detonated the breach charge.

The world rushed back into real time as the shaped charge decimated her foes. She was thrown backward from the blast. She knew that the only reason she survived the blast was because they were in a vacuum. She watched the ruined bodies fall to the ground, and then start to convulse as the air was pulled out of their suits.

Only one of the mech suits survived the blast, and charged forward with a speed that surprised Trish. She tried to get a shot off while she scrambled to her feet, but the pistol was kicked from her hand. A black armored hand shot out and grabbed her by the throat, then slamming her body so hard against a bulkhead that she saw stars in her vision. A man's face appeared before her when he changed his face plate from black to clear. The ice blue eyes bore into her. It was him. "I want to see your face," Ski growled on an open comm channel. "How about you go-" She was interrupted as her head slammed repeatedly against the wall until she almost lost consciousness.

She mentally kicked herself for changing her face plate to transparent, but she had learned long ago to cooperate and look for an opening in order to survive. "Well, I thought it was you. Sergeant, or is it Captain Keller now? Whatever. In a few minutes, it won't matter. WHERE. ARE. MY. CORE. BUSTERS?" He slammed her head with every word of his question. She was seeing stars again. Trish had to concentrate just form a sentence, "We destroyed them, sent 'em into a star a couple of jumps back." With an unhuman quickness, Ski pulled a small mono-blade from his suit and plunged it into Trish's ribs, puncturing her right lung. She fought the scream in her throat. Ski never took his eyes off hers as he twisted and then withdrew the blade. Her suit sealed the wound, but her head was swimming. She felt nauseous, and her lung was filling with blood. Her vision faded. She thought that she heard something, but it was so far away.

Chapter 45

Havoc felt the explosion through the walls and floor. He knew he was getting close. Havoc and Loki had been following the trail of bodies toward the engineering section of the ship for almost 20 minutes. His comm crackled to life. At first, the transmission was garbled and unreadable. As he continued on, he was able to make out a man talking. “WHERE. ARE. MY. CORE. BUSTERS?” Havoc almost stopped. It was Ski. “We destroyed them, sent ‘em into a star a couple of jumps back.” Havoc was sure that was Trish and she sounded hurt. Loki sprinted forward in reaction to the transmission. Loki disappeared around the corner with Havoc in the trail. As he cleared the corner, another comm came through the open net, “What the- AAHHHHH!”

Havoc stepped into the corridor and saw Loki’s helmet retracted. The armored dog was attached to Ski’s arm that was holding Trish to the bulkhead by the throat. Loki’s momentum carried all three of them onto the floor. Loki let go and turned to face Ski for another attack. Havoc had never seen Loki like this. His golden eyes were covered by a black membrane, his nostrils seem to have sealed shut, and the saliva in his mouth was boiling off in the vacuum. In his armed suit, Loki was now a terrifying sight.

Havoc continued to charge forward as Loki attacked again. This time, instead of biting onto an arm, Loki was met with mono-blade thrust into his right leg. The loss of use of his mechanical limb caused Loki to stumble and end up in Ski’s free hand. Ski used the hand around Loki’ helmet collar to push himself to his feet. Once standing, Ski turned to face Havoc, holding Loki at arm's length. When the pair locked eyes, Ski smiled. "Havoc, I hope that is you. I really want you to see this." With a speed that Havoc could barely track, Ski pulled the mono-blade out of Loki's leg and plunged it into his chest between two of the armored sections. Loki whined once over the comm's and went limp. Ski twisted the blade as he removed it and threw the dog's body into the bulkhead next to Havoc.

To Havoc, the whole exchange only took a blink of an eye. Loki's body had not even hit the ground when Havoc loosened his auto cannon on its maximum fire rate into Ski's mech suit. The shear volume of fire made Ski stagger back. An instant later, Havoc's larger mech suit smashed into Ski with enough force to kill an unarmored person. Even though Ski was a much larger man without a suit on, in the mech suit, Havoc stood eye level with him as the paired struggled to gain an advantage.

As the pair fought, Havoc had to remember to fall back onto his training. He pulled up the team's suit monitors while dodging a punch that would have cracked his face plate. He saw that Trish was in stable condition, just unconscious. He order her suit to inject a combat cocktail to get her back into the fight. Loki was in a far worse condition. Havoc landed his knee to Ski's midsection, which made him stagger back. Havoc pressed the advantage while sealing Loki's suit and putting him into stasis. Out of the corner of his vision, he saw Loki's helmet slam shut.

With the room he gained from the knee to Ski's midsection, Havoc pulled his own mono-blade from the built-in sheathe in his left arm. Ski took a step back and was immediately staggered by Havoc's auto cannon. Trish, who was just lying against the wall mere seconds ago, was now launching through the air with her mono-blade aimed at the soft armor on Ski's neck. She was just able to pierce the seal and draw blood before she was grabbed out of midair. Ski used her momentum and spun to throw her against the nearest bulkhead. Her body made a sickening sound as it collided with the wall. While his back was turned, Havoc took the opportunity to charge his mono-blade first. His target was the seam where Ski's suit closed. Havoc's blade sank deep into the suit and into Ski spine, Ski roared as his legs went limp.

Trish was on her feet as Ski fell in the middle of the deck. Blade in hand, she plunged her short sword through the front of his neck seal and pulled it across, cutting all the way into the spinal cord. Havoc stared into Ski's ice blue eyes. There was no panic, no fear, just hate. As his last act, Ski smiled and mouthed something that was just a garbling sound over the comm as blood drained from his neck and boiled away. Havoc stood frozen, watching Ski's body twitch for a few seconds. Trish snapped him out of his thoughts, "We have to get Loki out of here."

BD's voice came over the comm. "Captain, get out now. The XS-12 on the asteroid will go active in 3 minutes. You need to take Ski's shuttle and get out of there now." Without another word, Havoc grabbed Loki's limp body and sprinted after Trish.

BD had cracked the enemy's communication encryption. She now had access to all of their comms and, in a few seconds, would have access to all of their navigation data as well. Within three seconds, she had read only access to their entire system. After their attack run at her, they were going to spread out in the asteroid field and attack her from all sides. If she survived the upcoming missile attack, her simulations showed that she would not survive their next one.

BD saw in the navigational logs that Ski's shuttle had left Charlie 2 and docked with Bravo 4. The comm chatter indicated that Ski and his personal guards were engaged in a fire fight with at least one boarder, possibly two. She was able to check all suits vitals on the ship except for the Captain's, Loki's, and Havoc's. She worked to get camera feeds when nine suits redlined. The only one left was the mech suit that was identified as Ski's. She watched as his heart rate skyrocketed, and his suit indicated that he had just taken a dangerously high amount of an experimental combat cocktail.

Right before she plunged back into the chaff cloud that still lingered next to the remnants of Alpha squadron, she saw that Ski's heart rate flatlined. She updated her simulations, and a new course of action presented itself as having the highest probability of survival. She quickly sent an overpowered comm burst that would make hiding in the chaff cloud much harder, but it was a risk that she had to take. "Captain, get out now. The XS-12 on the asteroid will go active in 3 minutes. You need to take Ski's shuttle and get out of there now." The Beautiful Death plunged into the chaff cloud, and the comms dropped.

BD made a hard turn while in the chaff cloud to the still burning wreckage of Alpha squadron and cut her engines. Her calculations had the first missiles reaching her within the next ten seconds. She used her cold maneuvering jets to attach herself to the burning wreckage of Alpha 2. With her sensors blocked, she could not see the missiles streaking toward her. She only felt the vibrations from the impacts on Alpha 2.

The barrage lasted a full 30 seconds and, in that time, her shield was completely destroyed. She took two direct hits to her static shields. The first exploded and almost completely drained her shield. The second exploded and caused the shield to fail completely, the hot plasma and shrapnel from the missile tore through her cargo area and medical bays. The Beautiful Death was now open to the vacuum of space.

After a system test, BD determined that she would still be able to execute the plan with a 42.7% chance of everyone surviving. Now, she had to wait for Ski's shuttle to escape from Charlie squadron.

Chapter 46

Havoc sprinted after Trish with Loki cradled in his left arm. Havoc's mind was torn between worrying about Loki and following Trish through the corridors of the ship. Havoc only had to fire a couple of shots with his rifle. Trish and the auto cannon on his suit did the rest. As they ran, Havoc accessed Loki's suit controls to ensure that the stasis field was intact. Havoc had to keep his emotions in check when the health report came back with *"Critical condition. Seek emergency trauma care immediately. Chance of survival: Unknown"*. Havoc refocused and pressed on chasing Trish. When he did, he noticed something was wrong. She was slowing down and starting to miss shots. She never missed.

After clearing the corridor to the hangar, they saw Ski's shuttle in the almost completely hanger. The space craft did not look like anything Havoc had ever seen. It looked like a predator, completely blacked out with sleek and aggressive lines. Havoc shot two guards that Trish missed as she ran into the shuttle's open rear ramp. Havoc was just coming up the ramp when he saw the flash of Trish's pistol. She hurled the pilot's body into the cabin area, staining the plush carpet with the pilot's blood. "Strap in. BD, I need some help here. Can you take over controls for the shuttle? I am plugging my data cable. Use my suit as a bypass for the security protocols." To Havoc, she sounded like she was in trouble, her words were strained, and he could tell she was on the verge of passing out again. In less than a minute, they were clearing the hangar in Ski's state of the art shuttle craft.

BD had the shuttle's throttle maxed out as they headed toward the Beautiful Death. They were down to 30 seconds before they singularly started to pull all matter into it. The Beautiful Death appeared from the wreckage of Alpha squadron, her aft section was in ruins, and she was running on two of her four engines. BD maneuvered the shuttle into the aft section of the Beautiful Death, where the hangar bay used to be. The docking of the two ships was more of a controlled crash than any type of landing.

Charlie squadron was dispersing in the asteroid field all around the buried core buster. They had begun to make their hard turns back toward the fleeing Beautiful Death when the XS-12 unleashed the singularity. The entire section of the asteroid belt began to collapse into the miniature black hole. Charlie squadron was pulled in or smashed to pieces by the asteroids and mines that were being pulled into the miniature black hole at accelerating rates.

The Beautiful Death slowly started to lose velocity until she was at a fixed point in space, BD updated her simulations with the new information. The probability of all of their survival was approaching zero. Her prior simulations had at least three engines running at full power. Now, she had only two running at 83%. BD left the cockpit and traveled through the ruined ship until she reached the engineering spaces. "Captain Keller, I need you to authorize me to overpressure the engines."

"How are you going to do that? I thought that had to be done manually from inside the main reactors." Trish's voice sounded pained. BD checked the Captain's suit vitals; her right lung was completely filled with fluid, her heart rate was dangerously high, and her blood pressure was falling. The only thing keeping her conscious was the combat stim coursing through her blood.

"That is correct, Captain. But if I don't do this now, we will all die."

"What will happen to you when the reactor blows?"

"Captain, there is a chance I may expire, and you, Loki, and Havoc will live. Or if I do not do this, we all die in the next 20 seconds."

"Do it." Trish knew that BD had lied about there being a chance that the android would survive. Trish had destroyed enough reactors in her past life to know there was no way that BD would survive.

Behind the Beautiful Death, a bright flash appeared that briefly out shone the system's sun, the shock wave was coming. Havoc had heard the conversation between Trish and BD. He thought that BD might make it that maybe they all would. Havoc looked at the acceleration seat next to him, Loki's inert form was strapped in. "5 seconds to impact," BD called. Havoc gently grabbed Loki's paw and braced for the shock wave.

BD smiled, she was happy that her termination would mean that her friends had a chance to survive. She would miss Loki the most. She overloaded the reactor, and the ship shot forward just seconds before the blast wave hit. The reactor exploded less than two seconds later, completely destroying the front half of the Beautiful Death.

The engines continued to push what was left of the back half of the ship until they ran out of fuel.

Trish felt the violent explosion, and she was staring out into open space through the shuttles plazsteel windows. Just a second before she was looking into the ruined VR training room. The engines had given the rest of the ship just enough thrust not to be crushed by the destructive power of the XS-12. The shock completely overwhelmed the inertia dampeners, and both Trish and Havoc blacked out.

Havoc woke up to an eerie silence in the shuttle. He tried to blink to clear his eyes, but there seemed to be a red tint to everything. He checked the shuttle's atmosphere and found it was still breathable, so he opened his helmet to carefully rub his face. When he pulled his right glove away, it was covered in blood. Havoc laughed, he had no idea why. He slowly made his way to the cockpit. He was so tired, and his vision started to get blurry and then slowly come back into focus in a nauseating rhythm. He stumbled to the pilot's seat to look at Trish, the inside of her visor was covered in a red film. He wondered if she could see out of it. He was so tired, a flashing light in space caught his attention.

He tried so hard to think, the shape seemed so familiar. The situation rushed into focus, that was Gus's Folly. In a moment of clarity, Havoc saw that the shuttle was set to auto dock. He stumbled over to Trish, using his HUD to open her faceplate. Trish had bloody foam coming out of her mouth and nose. The foam was moving, so she was still breathing. The bump of a hard lock let Havoc know that the two ships were connected. He stumbled as he scooped up Trish and headed to the air lock at the rear of the ship. On his way through the cabin, he passed Loki's inert form pinned into the acceleration seat.

His heart broke. Havoc's two best friends lay dying in front of him. He was frozen with indecision. He stood stone still, unable to pick which one got to go into the med pod first. Blood stained tears splashed into Trish's face. Havoc looked down and heard Trish gasping for breaths, "I am so sorry, buddy. Please hang on." Havoc turned and stumbled through the airlock. The med pod was open and waiting. As soon as he entered the cargo area, through the fog in his mind, he was sure that he had not left it like that. Havoc opened up Trish's mech suit and he climbed out of his. He stripped Trish out of her mech suit as fast as he could, under her right breast was a triangle shaped wound that started to spurt bright red blood as soon as she was out of the suit.

Havoc rushed to place her in the med pod as fast as he could and slammed the lid closed. He stood there and watched as the alerts started flashing on its displays. The world started to spin. He could no longer keep the contents of his stomach from erupting from his nose and mouth. Havoc could no longer stand and fell to his hands and knees. Something was wrong. Havoc looked at what he just threw up, it was mostly blood. Havoc crawled to the first aid next to the med pod. He used the wall to climb up and reach up for the white and green box. Havoc managed to knock the box off the wall with a clumsy swipe. He followed it to the floor with a thump. The world spun faster. He opened the kit and fumbled through its contents until he found the nanite syringes. Havoc knew he was in bad shape, but hopefully, the medical innates would keep him alive until Trish was out of the med pod. He took a deep breath and jammed the auto syringes into his neck. The world went black before he was able to remove the needles from his body.

Epilogue

It was a full month before Gus's Folly was able to reach the next fringe station, as soon as he was able to hook into the space-net, he scoured the news to see what was being said about a fleet of ships disappearing or the massive explosion that occurred in the uninhabited system where the battle took place. There was nothing, not a single article, written about anything that resembled what had happened.

When was time to dock at the station, he headed back to the cargo bay to put on his mech suit. This place had a reputation for being a little rough. Havoc just wanted to pick up his supplies and head into the core worlds where he could refit and pick up a new ship. Havoc had decided that he was going to continue as a bounty hunter for a while, he need something to do while he tried to make sense of his life.

The bump of the hard dock let Havoc know that it was time to go out into the station. He closed his helmet, there was no reason to advertise who he was, just in case someone was still looking for him. "Come on, you two. I am not going to wait out here forever, and I am definitely not carrying back all of the supplies we need."

"Stop acting like you're in charge, you saved me once and now you think you run everything," Trish said as she exited the airlock. "Loki, hurry." Loki sprinted out of the ship, happy to be able to run at full speed again. Trish walked next to Havoc and lightly punched his armor with her bright pink fist. As they were walking down the concourse, Havoc and Trish got a plain text comm from an unknown user.

"WE NEED TO TALK. MEET US AT THE FOLLOWING COORDINATES."

Made in the USA
Middletown, DE
01 April 2025

73525702R00164